On The Trail of Blackbody Radiation

On the Trail of Blackbody Radiation

Max Planck and the Physics of His Era

Don S. Lemons, William R. Shanahan, and Louis J. Buchholtz

illustrated by Marta Gyeviki

The MIT Press
Cambridge, Massachusetts
London, England

The MIT Press would like to thank the anonymous peer reviewers who provided comments on drafts of this book. The generous work of academic experts is essential for establishing the authority and quality of our publications. We acknowledge with gratitude the contributions of these otherwise uncredited readers.

This book was set in ITC Stone Serif Std and ITC Stone Sans Std by New Best-set Typesetters Ltd. Printed and bound in the United States of America.

Library of Congress Cataloging-in-Publication Data

Names: Lemons, Don S. (Don Stephen), 1949- author. | Shanahan, William R., author. | Buchholtz, Louis J., author.
Title: On the trail of blackbody radiation : Max Planck and the physics of his era / Don S. Lemons, William R. Shanahan, and Louis J. Buchholtz ; Illustrated by Marta Gyeviki.
Description: Cambridge, Massachusetts : The MIT Press, [2022] | Includes bibliographical references and index.
Identifiers: LCCN 2021035122 | ISBN 9780262047043 (hardcover)
Subjects: LCSH: Blackbody radiation.
Classification: LCC QC484 .L46 2022 | DDC 530.12—dc23/eng/20211014
LC record available at https://lccn.loc.gov/2021035122

10 9 8 7 6 5 4 3 2 1

Dedications

Don S. Lemons: To his wife Allison and sons Nathan and Micah

William R. Shanahan: To his wife Katie

Louis J. Buchholtz: To his wife Barbara, daughter Clara, and son Will

Contents

Preface

It has been remarked that Max Planck chose, from among those systems in which the need for a quantum hypothesis is clear, the most complex of all: blackbody radiation. Planck spent years of strenuous effort bringing the latest tools of electromagnetism, thermodynamics, and statistical mechanics to bear on the problem of blackbody radiation. If only Planck had focused, as did Einstein in 1907, on the problem of the specific heat of a crystalline solid at low temperature, he might have produced a short, clearly written paper with groundbreaking implications.[1]

If, from one point of view, Planck chose the wrong problem, this was not a mistake. After all, Planck did not set out to usher in the quantum revolution, but rather to study a universal phenomenon for which empirical evidence had been accumulating since the late 1850s—that of blackbody radiation—a phenomenon that cried out for explanation. Planck knew that the radiation emitted and absorbed by nonreflective bodies in thermal equilibrium with one another is a universal phenomenon because its properties do not depend upon the physical characteristics of the emitting and absorbing bodies that produce it.

Formerly, students of physics studied the historical development of the theory of blackbody radiation and the early interpretations of that theory.[2] Such seems no longer the case. The dependence of the spectral energy density of blackbody radiation on temperature and wavelength is more often than not presented as an accomplished fact and its derivation, if given in modern terms, is reduced to a counting problem.

Even so, students have an interest in the very ingenious and human thinking upon which the quantum revolution in physics was founded. We know, for we are among those students. We still want to make sense of what Planck and his contemporaries thought about blackbody radiation. We want not only to verbally express those thoughts but also to work out their mathematical expression. We want to understand how Planck deployed his considerable knowledge of contemporary physics and to understand how Einstein and others used and interpreted Planck's work.

We found that while following the trail of Planck's thinking it was necessary to construct a kind of physics textbook composed of the late nineteenth-century concepts from which Planck drew while fashioning his derivation of the spectral energy density of blackbody radiation. Fortunately, the contents of this textbook are still current in classical mechanics, electromagnetism, thermodynamics, and statistical mechanics courses. What we have done is organize that content as a narrative of Planck's discovery. This narrative has become *On the Trail of Blackbody Radiation*. Thus, this work is partly historical, with appropriate reference to primary sources, and partly tutorial, with a translation of these contents into a scientific and mathematical language accessible to us and other students of physics.

In determining how much of Planck's thinking and its historical context to present we have tried to keep the needs of students foremost in mind. For this reason, sometimes we have accompanied an original derivation with others that to us seem clearer, while making reference to parallel and contrasting features in the original. And sometimes we have generously interpreted Planck's derivations while attempting not to falsify his approach.

For the historical context of blackbody radiation we have made significant use of Thomas Kuhn's *Black-Body Theory and the Quantum Discontinuity, 1894–1912.*[3] Primary sources include existing English translations of Planck and Einstein's papers. We have also translated two seminal papers in blackbody studies, we believe for the first time, from their original German into English: Ludwig Boltzmann's 1884 derivation of the Stefan-Boltzmann law and Wilhelm Wien's 1893 derivation of the Wien displacement law.

No doubt there are a number of relevant primary and important secondary sources that we have not consulted. Even so, studying the core sources has been a delight—for how ideas emerge can be as exciting as the ideas themselves. Part of our delight has been in achieving (what seems to us) new insight into the thinking of Boltzmann, Wien, Planck, and Einstein as they grappled with the problem of blackbody radiation—some aspects of which might be of interest not only to students of physics, but also to historians of science. These insights, briefly referenced in the two sections that bookend the text, "A Brief Guide to the Trail" and "The Big Ideas," are fully developed in the main text of *On the Trail of Blackbody Radiation.*

A Brief Guide to the Trail

The principle behind the composition of *On the Trail of Blackbody Radiation* is to include those concepts and methods upon which Planck drew in 1900–1901 while constructing his theory of blackbody radiation, and Einstein's reaction to Planck's theory, but not much else. And because these contributions are arranged in rough chronological order of their discovery, the text does not unfold as smoothly as would a typical pedagogically oriented text. Consequently, a reader who starts with chapter 1 and reads continuously through to chapter 12 may experience a series of intellectual jolts, as he or she passes from the merely verbal (chapter 1, "Prehistory"), to the seemingly mundane (chapter 2, "Classical Thermodynamics"), to the verbally framed if conceptually demanding (chapter 3, "Kirchhoff's Law"), and then on to the more mathematically oriented material in chapters 4 through 12 that even includes a counterfactual argument (section 8.3). This brief guide may help prepare readers for this jolting journey and also for appreciating what we believe are new insights into Boltzmann's, Wien's, Planck's, and Einstein's contributions.

Chapters 4 and 5 take up standard and not so standard derivations of the Stefan-Boltzmann law and Wien's displacement law. The not so standard derivations exploit the concept of adiabatic invariance—a concept that, to our surprise, plays a key role in Boltzmann's 1884 and Wien's 1893 presentations of their laws (see Appendices A and B for English translations). Today the concept of adiabatic invariance has a number of overlapping meanings that are not always distinguished from one another. Accordingly, we take pains, in chapters 2, 4, and 5, to clearly define what kind of adiabatic invariance Boltzmann, Wien, and we exploit.

Chapter 6, on the damped, driven, harmonic oscillator, may be the most familiar material in the text and, for this reason, might be quickly scanned. Yet be forewarned, the purpose of chapter 6 is to prepare the reader for the more mathematically demanding, electromagnetically damped and driven, Hertzian harmonic oscillator or "Planck resonator" as presented in chapter 7. The latter supplies the reasoning behind Planck's "fundamental relation."

Neither Wien's displacement law (chapter 5) nor the fundamental relation (chapter 7) follows straightforwardly from well-known principles of physics. However, anyone seeking to understand Planck's work and its early applications should understand both. For both are assumed as givens in Planck's two derivations of the spectral energy density of blackbody radiation (chapters 8 and 10) and in Einstein's early response to Planck's theory (chapter 11). Einstein's argument in his "Quantum Theory of Radiation" (chapter 12) also depends upon Wien's displacement law.

The mistake Planck made, in early 1900, when deriving Wien's (incorrect) distribution of the spectral energy density of blackbody radiation (distinct from Wien's displacement law) is identified in section 8.3. On the other hand, Planck's "zeroth derivation" (section 8.4) is sometimes dismissed as "mere curve-fitting." If so, its result has withstood the test of time. And this result gave Planck a goal at which to aim with more trustworthy methods.

When beginning his study of blackbody radiation in 1894 Planck believed the physical world was deterministic. Even so, he eventually adopted Boltzmann's statistical (and therefore probabilistic) characterization of entropy (chapter 9). Interestingly, Boltzmann's pioneering paper of 1877 also provided Planck with a ready-made mathematical path to the latter's desired formulation of the entropy of a Planck oscillator in a radiation field (section 9.4)—a path that Planck could have taken but did not, for, while adopting Boltzmann's statistical definition of the entropy, Planck employed a superficially distinct, if physically equivalent, combinatoric formulation of the entropy (section 10.3). In this way, Wien's displacement law, Planck's fundamental relation, Boltzmann's entropy, and Planck's combinatorics supplied the key elements out of which Planck composed his "first derivation" of the spectral energy density of blackbody radiation (chapter 10).

Aspects of Einstein's early response to Planck's derivation (chapter 11) are well known, as is the first part of Einstein's 1917 paper on the "Quantum Theory of Radiation" (chapter 12). They are included in the text because they shed light on and raise a question about the meaning of blackbody radiation. In particular, those who have grasped the apparent classical nature of Planck's derivation, as presented in chapters 5, 7, 8, and 10, will recognize that all the inputs to Einstein's theory of quantum radiation have classical derivations or classical analogs. And, while we can identify the point at which Planck (unintentionally) introduced the quantum into his derivation, Einstein's quantum hypothesis in his "Quantum Theory of Radiation" does not similarly announce itself. *On the Trail of Blackbody Radiation* concludes by noting this curious deficiency in Einstein's otherwise masterly derivation.

1 The Prehistory of Blackbody Radiation

1.1 Pictet's Experiment and Prevost's Exchanges

Among the early experiments that stimulated thinking about heating and cooling, none were more influential than *Pictet's experiment.* In 1790 Marc-Auguste Pictet (1751–1839) of Geneva reported that when he placed a thermometer at the focus of a metallic mirror aligned with another metallic mirror approximately 10 feet distant at whose focus was a flask of ice, "the thermometer . . . descended several degrees."[1] According to Pictet, "The fact was notorious, and amazed me at first; a moment's reflection, however, explained it. This phenomenon offered nothing more than a final proof, if it had been necessary, of the reflection of heat." Pictet's experiment is illustrated in Figure 1.1.

While an obvious, if superficial, interpretation of Pictet's experiment was that "cold" radiated in the same way as heat, it was not Pictet but rather Count Rumford (1753–1814), an itinerant American scientist known early in life as Benjamin Thompson, who jumped to this conclusion. According to Rumford, "frigorific rays" emanated from the ice, reflected from the metallic mirror at whose focus the ice sits, propagated

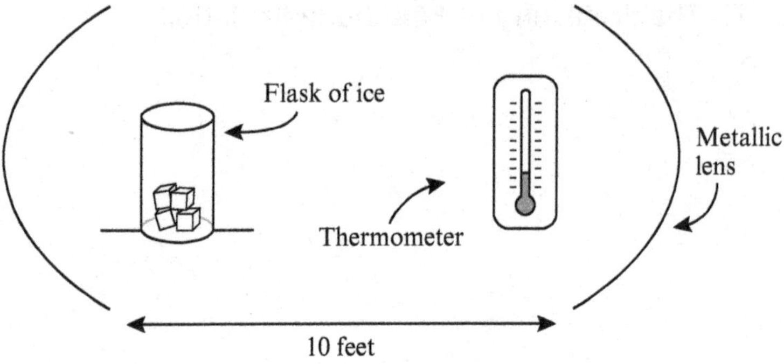

Figure 1.1
Pictet's experiment.

to the mirror with which it is aligned, and focused on the thermometer whose temperature subsequently "descended several degrees." Rumford dismissed other explanations of Pictet's experiment because he held that it is only the absorption of either calorific or frigorific rays, and not their emission, that changes the temperature of a body.

However, Rumford's interpretation has its problems. For instance, it must have been known in Rumford's time that sea ice was several degrees colder than freshwater ice. So, what would happen if sea ice were placed at one of the foci of Pictet's lenses and freshwater ice at the other? Would the freshwater ice suddenly stop emitting frigorific rays and start absorbing them? And, if so, at what temperature do objects quit emitting frigorific rays and begin emitting calorific rays?

Pierre Prevost (1764–1823), Pictet's colleague at the Geneva Academy, an educational institution founded by John Calvin in 1559, had an explanation that avoided these troubling questions. According to Prevost, both the thermometer and the ice in Pictet's experiment emit, absorb, and in this way exchange calorific rays. Furthermore, the higher a body's temperature, the more *caloric* or heat it emits. Because Pictet's thermometer emitted more and absorbed less heat than the ice, its temperature fell when it and the ice were placed at the foci of the two aligned metallic mirrors.

Prevost's interpretation became known as *Prevost's theory of exchanges*. According to this theory, all objects exchange calorific rays. Frigorific rays are no more than relatively weak calorific rays. Even two objects in thermal equilibrium, that is, two objects with the same temperature, exchange calorific rays, each one receiving as much heat as it loses. Thermal equilibrium is a dynamic rather than a static relation.

Rumford and Prevost not only envisioned the transfer of heat and cold in different ways, they held different views of their nature. For Rumford calorific and frigorific rays were undulations in a surrounding, invisible, elastic medium, while Prevost held that calorific rays are composed of particles of heat. Pictet favored Prevost's explanation, but many of their contemporaries remained undecided. Eventually, both Prevost's theory of exchanges and something akin to Rumford's undulations became part of a widely accepted explanation.

But just as important as the dispute over Pictet's experiment is what was not disputed. For it was clear to Pictet and to all concerned that something traveled along straight lines between Pictet's two metallic mirrors and reflected from them in much the same way as visible light travels between and reflects from mirrors. Thus, heat (and, for Rumford, cold) is *radiant* in the same way that light is radiant. That heat (and, for Rumford, cold) reflects from some materials (Pictet's metal mirrors) and is absorbed by others (his thermometer) was also beyond dispute.

1.2 Reflectors, Absorbers, and Emitters of Radiant Heat

Rumford and Prevost, both meticulous experimenters, also addressed the important question of what makes a good absorber or emitter of heat. Rumford's purpose was practical: he wished to determine the insulating properties of various materials in order that the Bavarian army, for whom he worked during the period 1785–1798, would be better protected from cold weather.

In pursuit of his goal Rumford filled two identically constructed metal containers with hot water. The surface of one container was highly polished while the other was successively covered with soot, varnish, glue, and linen. Rumford knew that the fur of animals kept them warm by inhibiting cooling by the convection of air. So he may have been surprised that the containers of hot water covered with layers of soot, varnish, glue, and linen cooled more quickly than the container with the highly polished surface.

Pierre Prevost arrived at similar conclusions as documented in his 1809 *Treatise on Radiant Heat.*[2] According to Prevost, "A reflecting body, having been heated or cooled internally, recovers the surrounding temperature more slowly than a non-reflector," and "A reflecting body, having been heated or cooled internally, will have less effect on another body placed at any distance [in heating or cooling it] than a non-reflector would under the same conditions."

As a general rule, surfaces that reflect well (that is, absorb poorly) maintain their interior heat well (that is, emit poorly.) And surfaces that absorb well, emit well. A modern application of this rule is the practice of first responders who carry blankets covered with reflective material. Since these blankets reflect well, they also emit poorly and, for this reason, help maintain the body heat of accident victims who are out of doors in cool weather. Ordinary blankets are constructed differently, for indoors the unequal radiation of heat is not so important a cause of cooling as out of doors.

1.3 Blackbodies and Blackbody Radiation

It only remained to determine that the radiant heating and cooling Prevost and Rumford observed are caused, respectively, by the absorption and emission of waves that include the waves of visible light for the concept of a *blackbody* to be formulated. Indeed, in 1800 Thomas Young (1773–1829) first demonstrated that light has wave properties and, in the same year, William Herschel (1738–1822) discovered an invisible continuation of the spectrum of sunlight beyond its red end that heats a thermometer as well as or better than visible sunlight. (Herschel's discovery of *infrared* light is illustrated in Figure 1.2.) Both discoveries initiated research that by the mid-1800s established that heat is carried by waves much like visible light.

Since bodies with surfaces that absorb visible light and radiate mainly in nonvisible parts of the spectrum appear to us as black, the word *blackbody* has been adopted, somewhat misleadingly, to stand for all objects that are good absorbers of *radiation,* that is, good absorbers of light. For we have learned that good absorbers are also good emitters. And some objects that are good absorbers of visible light also radiate in visible parts of the spectrum. Thus, blackbodies don't always appear black.

Consider, for instance, the coils of an electric burner as they begin to radiate heat. The color of the coils, originally a dull black, at first remains unchanged even as the coils become too hot to touch. Then, as more heat is produced, the coils acquire

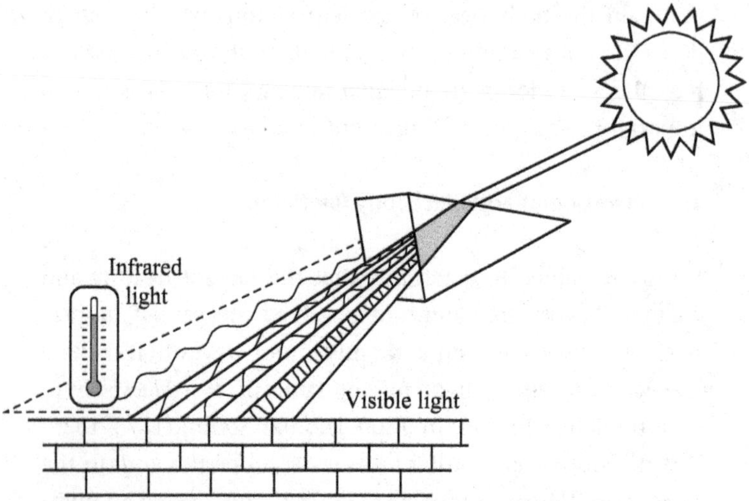

Figure 1.2.
William Herschel discovered the heating power of infrared light in 1800.

a reddish tinge and finally at maximum heat are bright red. Similar changes occur in the color of the coals in our fireplaces and wood-burning stoves. As the coals become hotter their color becomes brighter. The coils of an electric burner and the coals in our fireplaces and wood-burning stoves are approximate blackbodies. And the correlation of their temperature with their color is characteristic of the kind of heat production and heat transfer we call *blackbody radiation.*

To summarize, *radiation,* in the phrase *blackbody radiation,* refers to heat transferred by the spectrum of waves of which light is a part, *blackbody* stands for an object that absorbs all the radiation falling upon it, and definite color changes correlate with the temperature of a blackbody.

2 Classical Thermodynamics

2.1 Why Thermodynamics?

The answer to this question lies in the high regard that Max Planck and Albert Einstein had for the thermodynamics of their day—what we now call *classical thermodynamics*. Both saw thermodynamics as the *beau idéal* of a physical theory. As a young man Planck wrote a thesis and a dissertation on the second law of thermodynamics and, in 1897 in the midst of his investigation of blackbody radiation, the mature Planck published a textbook on the subject, *Treatise on Thermodynamics*.[1] Einstein famously claimed that thermodynamics "is the only physical theory of universal content concerning which I am convinced that, within the framework of applicability of its basic concepts, it will never be overthrown."[2] Both men, in their theorizing, were strongly influenced by the example of classical thermodynamics.

But we have more specific reasons for inserting a chapter on classical thermodynamics into this narrative, for thermodynamic concepts and arguments provide the framework for Planck's investigation of blackbody radiation. One has only to notice the many references in later chapters to the concepts of *entropy*, *temperature*, and *adiabatic invariance*. In chapter 3 we use the second law of thermodynamics to prove Kirchhoff's law, and in chapters 4 and 5 we expose the way in which Ludwig Boltzmann and Wilhelm Wien exploited the adiabatic invariants of a system of blackbody radiation to derive the original versions of their eponymous laws. In chapter 9 we present Boltzmann's probabilistic formulation of the entropy. Planck's subsequent work, as presented in chapters 7, 8, and 10, builds directly upon that of Kirchhoff, Boltzmann, and Wien.

While most readers will have some acquaintance with classical thermodynamics, grasping its intellectual structure to the extent necessary to fully appreciate Planck's and Einstein's work is another matter. Accordingly, some readers for whom classical thermodynamics is well known may skip this chapter. Others will want to review it.

2.2 Equilibrium and the Zeroth Law of Thermodynamics

Rudolf Clausius (1822–1888) formulated the structure of classical thermodynamics between 1850 and 1865, publishing a series of nine memoirs on the subject.[3] Clausius's thermodynamics, which identifies the first and second laws of thermodynamics, and Boltzmann's 1877 statistical formulation of the entropy of a system are the foundations upon which Max Planck built his theory of blackbody radiation in 1900–1901.

However, Planck's theory of blackbody radiation and, indeed, all applications of thermodynamics depend upon a principle that, while implicit in Clausius's formulations, was made explicit only around 1935.[4] We now call this principle *the zeroth law of thermodynamics* according to which *two thermodynamic systems each in thermal equilibrium with a third are in thermal equilibrium with each other*. For this reason, the zeroth law depends upon the concepts of *thermodynamic system* and *thermal equilibrium.*

A thermodynamic system is a macroscopic part of the universe, like a bucket of water or a piece of rock, on which we focus our attention. Of course, the physical state of such a system may be quite complex. For the water in the bucket may be turbulent and at any given instant a shock wave may be propagating through the rock. But, interestingly, when left to themselves for a period of time, macroscopic systems always settle down to a quiescent state, an *equilibrium state,* that can be described, simply and fruitfully, with a mere handful of *state variables.* Similarly, two systems in thermal contact will after a period of time achieve *mutual thermodynamic equilibrium* in which the state variables describing each system do not change.

From among thermodynamic systems a particular one can be chosen for the purpose of using its state variables to define a particular state variable called the *empirical temperature.* For instance, the volume of a column of mercury in a glass tube is a traditional indicator of empirical temperature. An empirical temperature tells us when two systems are in thermal equilibrium.

The coherence and utility of the concept of empirical temperature depend only upon the zeroth law of thermodynamics. (*Absolute* or *thermodynamic temperature* is another concept whose coherence and utility depend on both the first and second laws.)

While Clausius took the concept of empirical temperature for granted, it is foundational to the structure of classical thermodynamics he erected. In the following sections we outline that structure, including the first and second laws of thermodynamics. We then apply these laws to the ideal gas system. In chapters 3 and 4 we take up the classical thermodynamics of blackbody radiation.

2.3 The First Law of Thermodynamics

There are two ways to change the equilibrium state of a system: (1) performing work on that system or arranging for that system to do work on its environment and (2) manipulating its environment to heat or cool the system. According to the first law of thermodynamics, the work W done on a system and the heat Q transferred to the system accomplish equivalent changes in the state of the system. For instance, both work and heat can raise the temperature of a system. Indeed, the founders of thermodynamics more often spoke of the *principle of the equivalence of work and heat* than they did of the *first law of thermodynamics*.

Accordingly, work W and heat Q can be denominated in the same units. Furthermore, it is the sum $Q + W$ of these two quantities that changes the state of that system. For this reason, we may denote the first law as

$$\Delta X = Q + W \tag{2.1}$$

where ΔX stands for a change in a quantity X (soon to be renamed) whose only purpose here is to quantify the change in the system's state caused by changes in the sum $Q + W$. Changes in X are correlated with changes in other state variables, such as volume V and pressure P, through the system's *equations of state*.

Note the difference between the variable X and the variables Q and W. The former is, by design, a state variable, while the latter are different ways of contributing to a sum that changes that state variable. In other words, a system is characterized by X but not by Q or W. In no way does a system possess either Q or W. But a system may be said to possess X.

When the heat transferred to and the work done on a system are differentially small, that is, are δQ and δW, the change dX to the state variable X is also differentially small. Then (2.1) becomes

$$dX = \delta Q + \delta W \tag{2.2}$$

where the different notations, dX and δQ or δW, are meant to reinforce the difference between a state variable X and different ways of changing that state variable δQ and δW.

When the system is an isotropic fluid, characterized by an isotropic pressure P, the differential work δW done *reversibly* on the fluid is given by $\delta W = -PdV$. *Reversible work* is that done without friction or internal dissipation and indefinitely slowly—in a word, *quasistatically*. The negative sign means that when the fluid expands, so that $dV > 0$, negative work $-PdV$ is done *on the fluid system* and positive work PdV is done *by the fluid system*. When $-PdV$ work is the only kind of work done on the system, the constraint (2.2) becomes

$$dX = \delta Q - PdV. \tag{2.3}$$

Yet one more notational change will bring (2.3) closer to a familiar version of the fundamental constraint on a fluid system. If we call the state variable X by the name *internal energy* and use Clausius's symbol U in its place, equation (2.3) becomes

$$dU = \delta Q - PdV. \tag{2.4}$$

Equation (2.4) is the differential form of the *first law of thermodynamics* for a system of isotropic fluid.

2.4 Thermodynamic Temperature

The first law of thermodynamics allowed Clausius to update Sadi Carnot's (1796–1832) earlier concept of a perfect heat engine or *Carnot engine,* that is, one in which work is done and heat transferred *reversibly* or, equivalently, quasistatically and without friction or dissipation. Carnot had introduced the concept of a perfect heat engine in his essay *Reflections on the Motive Power of Fire* (1824) in order to prove several theorems concerning the optimal operation of a heat engine.[5]

Interestingly, Carnot mistakenly believed that heat rather than energy was conserved. Therefore, according to Carnot, a perfect heat engine produces work by allowing heat to flow, without diminution or increase, from a hotter heat reservoir through the heat engine to a colder one much as a quantity of water is conserved as it performs work by flowing over a mill wheel or by flowing through the turbine of a hydroelectric generator. Figure 2.1 diagrams the energy flow through one cycle, a *Carnot cycle,* of a Carnot engine as modified by Clausius to conserve energy as required by the first law.

William Thomson (1824–1907), known late in life as Lord Kelvin, used Clausius's version of Carnot's perfect heat engine to define another ideal—that of an *absolute* or *thermodynamic temperature.*[6] According to Thomson, the thermodynamic temperature T_H of the hotter heat reservoir of a Carnot engine is, by definition, proportional to the heat Q_H that is extracted from it during one cycle of the engine and, similarly, the thermodynamic temperature T_C of the colder reservoir of a Carnot engine is proportional to the heat Q_C that it receives during one cycle of the engine. Therefore, by his definition, $T_H \propto Q_H$ and $T_C \propto Q_C$ or, equivalently,

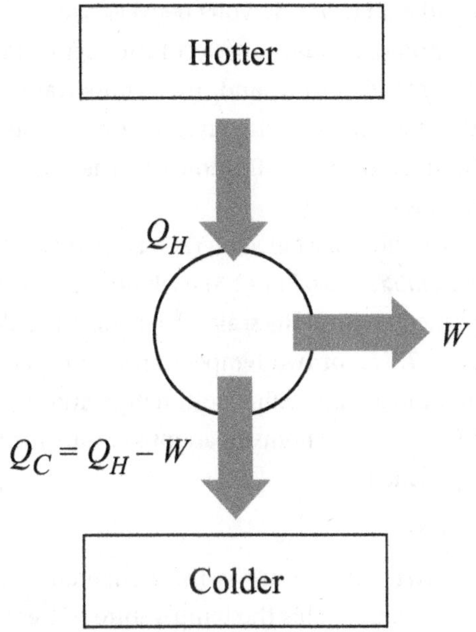

Figure 2.1.
Energy flow through an energy-conserving, perfect heat engine operating in a Carnot cycle. In one reversible cycle the engine extracts heat Q_H from a hotter heat reservoir, produces work W, and expels waste heat $Q_C = Q_H - W$ to a colder heat reservoir.

$$\frac{T_C}{T_H} = \frac{Q_C}{Q_H}. \tag{2.5}$$

Thomson's relation (2.5), the definition W/Q_H of the efficiency of a heat engine, and the conservation of energy $W = Q_H - Q_C$ together imply that

$$\frac{W}{Q_H} = 1 - \frac{T_C}{T_H}, \tag{2.6}$$

which expresses the *Carnot efficiency*, that is, the efficiency of a Carnot engine in terms of the thermodynamic temperatures of the hot and cold heat reservoirs between which it operates. Of course, no one can build a perfect heat engine. But one can build approximations to it, and, more importantly, one can measure thermodynamic temperatures by measuring quantities deduced from their definition and the laws of classical thermodynamics.

Any temperature scale based on (2.5) is an *absolute or thermodynamic temperature scale.* But (2.5) itself does not by itself fully define a particular absolute scale. After all, (2.5) determines only the ratio T_C/T_H of two temperatures. Another condition is needed to individually define both temperatures T_H and T_C.

In addition to (2.5) Thomson adopted a rule for the difference $T_H - T_C$, namely,

$$T_H - T_C = 100°, \tag{2.7}$$

when the hotter temperature is the temperature of boiling water and the colder one is the temperature of freezing water. Thomson's rule (2.7) made the size of an absolute degree equal to that of the Centigrade scale then in use. Together (2.5) and (2.7) define the *Kelvin scale* and the size of a *degree Kelvin.*

Other absolute scales with other rules in place of (2.7) are possible. For instance, the *Rankine scale* is defined by (2.5) and the rule $T_H - T_C = 180°$ for the temperature difference between boiling and freezing water. In this way, a *Rankine degree* is equal to a Fahrenheit degree.

2.5 The Second Law of Thermodynamics

All thermodynamic temperatures defined by (2.5) lead to a result

$$\frac{Q_H}{T_H} - \frac{Q_C}{T_C} = 0 \tag{2.8}$$

that can be interpreted as the statement that, during one cycle, a Carnot engine receives a quantity Q_H/T_H from the hotter reservoir and expels another quantity Q_C/T_C, with the same numerical value, to the colder reservoir. Equation (2.8) is, indeed, a special case of a more general result called *Clausius's theorem* that establishes the existence of a new state variable.[7] During incomplete cycles of a Carnot engine the values of Q_H/T_H and Q_C/T_C are not the same. In these cases, the value of this new state variable is changed. Clausius chose to call this new state variable the *entropy* and to denote it with the symbol S.

Clausius' proof of his theorem starts with the first law of thermodynamics and his version of the second law, namely, that *no process whose only result is to transfer heat from a colder reservoir to a hotter one is possible*. From these hypotheses Clausius proved that the differential increment dS to the entropy state variable S made during the reversible heating ($\delta Q > 0$) or cooling ($\delta Q < 0$) of a system at temperature T is

$$dS = \frac{\delta Q}{T}. \tag{2.9}$$

Given the widespread tendency to associate entropy with the second law of thermodynamics, one cannot emphasize too much that (2.9) and, indeed, all statements about entropy are deductions from both the first and second laws of thermodynamics as formulated by the founders of classical thermodynamics.

Equation (2.9) transforms the first law as expressed in (2.4) into the *fundamental constraint*

$$dU = TdS - PdV \tag{2.10}$$

for a fluid system. This constraint restricts how the state variables, U, T, S, P, and V, of an isotropic fluid can be related to one another in equations of state, and, indeed, can help one transform incomplete equations of state into more complete ones—as we will show in the next two sections. The fundamental constraint (2.10) encapsulates the restrictions of the first and second laws of thermodynamics in terms of William Thomson's definition of the *absolute* or *thermodynamic temperature* T.

The fundamental constraint, which for an isotropic fluid is expressed by (2.10), and the laws of thermodynamics form a category of truths that are formally different from the category containing a system's equations of state. The former apply to all thermodynamic systems, while the latter are peculiar to and must be discovered individually for each kind of system.

2.6 The Fluid System

The fundamental constraint (2.10) of a fluid system is equivalent to an expression,

$$dS = \frac{1}{T}dU + \frac{P}{T}dV, \tag{2.11}$$

whose form determines the partial derivatives of the entropy function $S(U,V)$. For, given the functional dependence $S(U,V)$, the differentials dS, dU, and dV are related to each other by the identity

$$dS = \left(\frac{\partial S}{\partial U}\right)_V dU + \left(\frac{\partial S}{\partial V}\right)_U dV \tag{2.12}$$

where the notation $(\partial S/\partial U)_V$ indicates the partial derivative of $S(U,V)$ with respect to U while holding V constant. The fundamental constraint (2.11) for a fluid system and the identity (2.12) together require that

$$\left(\frac{\partial S}{\partial U}\right)_V = \frac{1}{T} \tag{2.13}$$

and

$$\left(\frac{\partial S}{\partial V}\right)_U = \frac{P}{T}. \tag{2.14}$$

These two expressions are formal versions of the two independent equations of state of the fluid system. For, if we know a system's entropy function $S(U,V)$, equations (2.13) and (2.14) determine its two independent equations of state. Alternatively, if we know both equations of state, we can integrate (2.13) and (2.14) and determine the system's entropy function $S(U,V)$. In this way, either the function $S(U,V)$ or the two independent equations of state tell us everything classical thermodynamics can tell us about a fluid system, whether that system is an ideal gas or blackbody radiation.

Furthermore, because the order of partial derivatives operating on the entropy function is, in thermodynamics, inconsequential,

$$\frac{\partial^2 S}{\partial V \partial U} = \frac{\partial^2 S}{\partial U \partial V}. \tag{2.15}$$

Accordingly, the formal equations of state, (2.13) and (2.14), imply through (2.15) that the state variables of a fluid system are constrained by

$$\frac{\partial}{\partial V}\left(\frac{1}{T}\right)_U = \frac{\partial}{\partial U}\left(\frac{P}{T}\right)_V. \tag{2.16}$$

This constraint is a consequence of those imposed by the first and second laws of thermodynamics.

2.7 Example: The Ideal Gas

Consider, for instance, Boyle's law in the form

$$PV = f(T) \tag{2.17}$$

where $f(T)$ is an unknown function of the thermodynamic temperature T of the gas. Experiments in which a given quantity of gas is allowed to expand into a vacuum without performing work or absorbing heat, called *Joule expansion*, provide evidence that the internal energy U of the gas depends only upon its thermodynamic temperature T and not upon its volume V. Indeed, the internal energy U of the gas is another unknown function $U(T)$. How do the first and second laws of thermodynamics constrain the unknown functions $f(T)$ and $U(T)$?

Given that the gas energy U is only a function of its temperature T, it must be that $(\partial T/\partial V)_U = 0$, since if the internal energy U of the gas is held constant, its thermodynamic temperature T is also held constant. Therefore, the left-hand side of constraint (2.16) vanishes, and the constraint reduces to

$$0 = \frac{\partial}{\partial U}\left(\frac{P}{T}\right)_V. \tag{2.18}$$

Given Boyles's law (2.17), this constraint is equivalent to

$$\frac{\partial}{\partial U}\left[\frac{f(T)}{T}\right]_V = 0, \tag{2.19}$$

which, in turn, is equivalent to the total derivative

$$\frac{d}{dU}\left[\frac{f(T)}{T}\right] = 0. \tag{2.20}$$

Equation (2.20) implies that $f(T)/T$ is a constant independent of both V and U. Therefore, $f(T) \propto T$, and Boyle's law becomes the proportionality

$$PV \propto T. \tag{2.21}$$

When the proportionality constant is nR where n is the quantity of gas in moles and R is the fundamental "gas constant," Boyle's law becomes

$$PV = nRT, \tag{2.22}$$

which is the familiar form of one of the two independent equations of state of an ideal gas.

Note, however, that the first and second laws of thermodynamics do not constrain the unknown function $U(T)$. A particularly simple realization of $U(T)$ is

$$U(T) = C_V T, \tag{2.23}$$

in which case C_V is called the *heat capacity* of the gas at constant volume. Gases that observe both (2.22) and (2.23) are called *ideal gases*.

2.8 The Adiabatic Invariant of an Ideal Gas

The fundamental constraint of a fluid system $dU = TdS - PdV$ describes the relationships among changes in the state variables U, T, S, P, and V allowed by the first and second laws of thermodynamics. When these changes are such as to preserve the system's entropy, so that $dS = 0$, the fundamental constraint reduces to

$$dU = -PdV. \tag{2.24}$$

Entropy-conserving or *isentropic* changes or processes described by (2.24) are, of necessity, reversible, and of course reversible processes are quasistatic, dissipationless ones. In addition to being reversible an isentropic process is one in which no heat $\delta Q[= TdS]$ enters or leaves the system. We call quantities that are invariant during isentropic processes *adiabatic invariants*.

The adiabatic invariant of an isotropic fluid solves the differential equation (2.24). But, of course, we can solve (2.24) directly only when we can express the pressure P in terms of the variables U and V. And, in general, the equations of state of the system must be known in order to determine $P(U,V)$. For example, expressions for the pressure $P[=nRT/V]$ and the energy $U[=C_vT]$ of an ideal gas allow us to determine that $P=(nR/C_V)(U/V)$. In this case, the entropy-conserving condition (2.24) becomes

$$dU=-\left(\frac{nR}{C_V}\right)\left(\frac{U}{V}\right)dV \tag{2.25}$$

which is easily solved to produce

$$UV^{nR/C_V}=ad.inv. \tag{2.26}$$

where "*ad.inv.*" stands for "adiabatic invariant."

We use the special symbol *ad.inv.* instead of *const.* to emphasize that an adiabatic invariant is constant only during an entropy-conserving process. We cannot, for instance, assume that an adiabatic invariant such as UV^{nR/C_v} is equal to a dimensionally correct combination of fundamental constants in the same way that we know that for an ideal gas $PV/nT=const.$ where here *const.* is an *absolute constant* called the *fundamental gas constant R.*

2.9 The Entropy of an Ideal Gas

The derivation of the adiabatic invariant of an ideal gas in section 2.8 implies that whenever the entropy of an ideal gas remains constant, its adiabatic invariant remains constant. But we know more than this. For, given the fundamental constraint $dU=TdS-PdV$ and the equations of state, $PV=nRT$ and $U=C_VT$, of an ideal gas, we also know that changes in the entropy are related to changes in the adiabatic invariant by

$$
\begin{aligned}
dS &= \frac{1}{T}dU + \frac{P}{T}dV \\
&= \frac{C_V}{U}dU + \frac{nR}{V}dV \\
&= C_V\left[d\ln U + \frac{nR}{C_V}d\ln V\right] \\
&= C_V\left[d\ln\left(UV^{nR/C_V}\right)\right].
\end{aligned} \tag{2.27}
$$

Therefore, we know that *the entropy of an ideal gas changes if and only if its adiabatic invariant changes.*

For this reason the entropy of a fluid system is a function of its adiabatic invariant. In particular, the dependence of the entropy of an ideal gas on its adiabatic invariant is made explicit by the functional form

$$S(U,V) = S\left(UV^{nR/C_V}\right). \tag{2.28}$$

Of course, the expression on the left-hand side of (2.28) is a function of two independent variables, while the expression on the right-hand side is, necessarily, a different function of only one variable. In this way, expression (2.28) exploits the concept of adiabatic invariance in order to reduce the number of variables on which the entropy of an ideal gas depends from two to one—a reduction that will prove useful to us.

Integrating (2.27) allows us to produce the form of the function in (2.28), namely,

$$S_2(U,V) = C_V \ln\left(UV^{nR/C_V}\right) + const., \tag{2.29}$$

where here *const.* stands for a function of relevant fundamental constants such as R and constants of this system, such as mole number n and heat capacity C_V. Derivatives of this expression for the entropy with respect to U and V and the formal equations of state, (2.13) and (2.14), generate the system's two independent equations of state, $U = C_V T$ and $PV = nRT$.

2.10 Relations among Different Forms of the Adiabatic Invariant

The adiabatic invariant of an ideal gas UV^{nR/C_V} can, with the help of the equations of state, $PV = nRT$ and $U = C_V T$, be cast into different forms. For instance, TV^{nR/C_V} and PV^{1+nR/C_V} are two of the several different forms of the adiabatic invariant of an ideal gas. *Since each form of the adiabatic invariant (including the entropy) changes if and only if every other form changes, each must be a function of each.* Furthermore, *because the different forms of the adiabatic invariant of a fluid system are related to each other through the equations of state, every function relating two different forms of the adiabatic invariant of a system reveals something of the content of the system's equations of state.*

For example, when the system is an ideal gas, the system's thermodynamic variables *U*, *V*, and *P* are related to each other by the following relation between different forms of its adiabatic invariant,

$$UV^{nR/C_V} = f\left(PV^{1+nR/C_V}\right) \tag{2.30}$$

where $f\left(PV^{1+nR/C_V}\right)$ is, as yet, an undetermined function that contains information about the equations of state of this system. However, since we already know the equations of state of an ideal gas, we can easily deduce the form this function must take, namely,

$$f\left(PV^{1+nR/C_V}\right) = \frac{C_V}{nR}\left(PV^{1+nR/C_V}\right) \tag{2.31}$$

where C_V/nR is a constant of the system. Equation (2.30) and the functional definition (2.31) together produce

$$UV^{nR/C_V} = \frac{C_V}{nR}\left(PV^{1+nR/C_V}\right), \tag{2.32}$$

which is equivalent to

$$U = \frac{C_V}{nR} PV, \tag{2.33}$$

a result that expresses part of the content of the two independent equations of state of an ideal gas.

3 Kirchhoff's Law, 1859

3.1 Blackbody Radiation and the Laws of Thermodynamics

The years 1850–1851 were a turning point in the history of physics that allowed questions concerning blackbody radiation to be meaningfully framed and answered. For it was in 1850 that Rudolf Clausius enunciated the two laws of classical thermodynamics. The first law made possible the quantification of a system's *energy*, while Clausius's version of the second law, *no process whose only result is to transfer heat from a colder reservoir to a hotter one is possible*, established a favored direction for processes. In 1851 William Thomson articulated his own version of the second law and used the first two laws of classical thermodynamics to form the concept of an *absolute* or *thermodynamic temperature*.

Here we take up the energy density and spectral energy density of blackbody radiation. We define the spectral energy density and the coefficients of absorption and emission of blackbody radiation. All are constrained by the laws of thermodynamics.

3.2 The Energy Density of Blackbody Radiation

Heat radiation passes through empty space at the finite speed of light. For this reason, a given volume of space at any one instant contains radiation. On what, then, does the energy density of this radiation depend when in thermal equilibrium with blackbody absorbers and emitters? Certainly this energy density depends, in some way, upon the thermodynamic temperature T of the object or objects producing the radiation, but on what else?

Our goal is to show that the energy density of blackbody radiation can depend only upon the thermodynamic temperature T of the bodies producing the radiation and not at all upon the volume or shape of the radiation or the composition of the radiating and absorbing bodies. Our proof of this statement is an indirect one that follows upon assuming the contrary statement, for instance, that the energy density of blackbody radiation depends upon the volume V of the radiation, and then showing that this assumption leads to a violation of one of the two laws of classical thermodynamics.

To implement our proof we imagine an isolated enclosure surrounded by adiabatic walls and separated into two chambers by a removable, adiabatic, rigid barrier that when in place prohibits energy transfer between the two chambers—as illustrated in Figure 3.1(a). The enclosure walls are also blackbodies and in thermal equilibrium with the radiation they enclose. Therefore, a single thermodynamic temperature T characterizes the walls and the radiation. Furthermore, the walls of the enclosure and the radiation it contains are finite objects with finite heat capacities. When the barrier is in place, the two chambers are separate enclosures, and when the barrier is removed as in Figure 3.1(b), the two chambers become one.

Initially, the barrier is in place and the enclosure walls are in thermal equilibrium with each other and with the blackbody radiation they contain as illustrated in Figure 3.1(a). Now, suppose that blackbody radiation energy density u does,

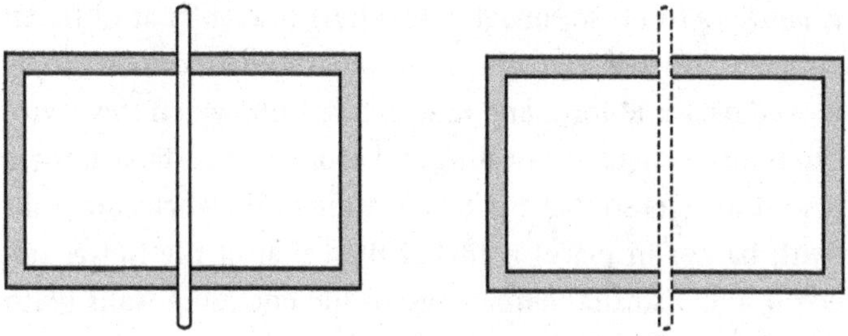

Figure 3.1.
(a) Two chambers of an adiabatic enclosure separated by an adiabatic, rigid barrier. (b) The enclosure with barrier removed.

indeed, depend upon the radiation volume V as well as on its temperature T so that $u = u(T,V)$. Furthermore, suppose that at a given constant temperature T, the larger the volume V, the larger the energy density u so that $(\partial u/\partial V)_T > 0$. Now we remove the barrier (as we suppose we can without performing work on the radiation), and the two chambers become one chamber with a larger volume. Then, according to our supposition, a new equilibrium is established with a larger blackbody radiation energy density u. This additional radiation energy can only come from the enclosure walls whose temperature consequently must decrease. Therefore, the new equilibrium temperature (with barrier removed) is less than the initial equilibrium temperature (with barrier in place). If the barrier were replaced, the energy density of the blackbody radiation in each part would decrease and the thermodynamic temperature of the enclosure walls would increase to its original value.

We will now use this extraordinary device to violate Clausius's version of the second law, namely, that *no process whose only result is to transfer heat from a colder reservoir to a hotter one is possible.* To do so, imagine two heat reservoirs at different temperatures between which the extraordinary device can be moved back and forth and to which or from which the device can transfer energy by heating and cooling. We establish these heat reservoirs so that the temperature of the enclosure walls (with barrier in place) is higher than that of the hotter reservoir and also the temperature of the enclosure walls (with barrier removed) is lower than that of the colder reservoir. Then, with barrier in place the enclosure walls can heat the hotter reservoir and with barrier removed the barrier walls can accept, from the colder reservoir, the heat the enclosure lost to the hot reservoir. Replacing the barrier restores the system to its initial condition, a process that violates Clausius's expression of the second law since its only result is to heat a hot reservoir and cool a cold one. Therefore, the initial supposition that the energy density of blackbody radiation depends upon volume must be false.

With a little imagination this device could be modified to show that neither can the blackbody radiation energy density u depend on the shape or the composition of the enclosure. For, in principle, both the shape and the composition of the enclosure could be changed without performing work on the radiation. Then, assuming that the energy density of blackbody radiation depends upon either the shape or composition of the enclosure, the same process of heating the hotter of two reservoirs and cooling the colder one similarly leads to a violation of Clausius's version of the second law. As a result, the energy density of blackbody radiation contained within a constant-temperature enclosure cannot be a function of enclosure volume, shape, or composition. The energy density of blackbody radiation is only a function $u(T)$ of the thermodynamic temperature T.

Of course, one could also imagine a heat transfer process that saves the second law, say, by stipulating that the temperature of the enclosure walls does not change when the energy density of the blackbody radiation changes. However, in this case, the first law would be violated. Therefore, the supposition that the energy density of blackbody radiation depends upon enclosure volume, shape, or composition must be false.

3.3 The Spectral Energy Density

The *spectral energy density of blackbody radiation* $u_\lambda(T,\lambda)$ is the blackbody energy density contained within a differential range of wavelengths λ to $\lambda + d\lambda$. According to this definition, the energy density u and the spectral energy density u_λ are related by

$$u(T) = \int_0^\infty u_\lambda(T,\lambda) d\lambda. \tag{3.1}$$

Since the energy density $u(T)$ of blackbody radiation is a function of only the thermodynamic temperature T, the spectral energy density $u_\lambda(T,\lambda)$ can only be, as indicated, a function of thermodynamic temperature T and wavelength λ.

3.4 Kirchhoff's Law of Thermal Radiation

Kirchhoff's law of thermal radiation, which we now present, is concerned with the degree to which different materials absorb and emit radiation. Its expression is in terms of a dimensionless material's absorption coefficient a that indicates the fraction of incident energy that is absorbed. Thus, when $a = 1$, a material absorbs all the radiation falling upon it, and when $a = 0$ the material absorbs none of the radiation falling upon it. Because radiation that is not absorbed is, in some way, reflected, the fraction of incident radiation reflected is $1 - a$. Kirchhoff's law also concerns a coefficient of emissive power e where the latter indicates the rate at which a material emits radiant energy from a unit surface area. Thus, the emissive power e has units of power per unit area.

The dimensional emissive power e and the dimensionless absorption coefficient a describe the emitting, absorbing, and reflecting properties of a particular material. One might reasonably expect that e and a depend upon the composition of the material—and separately they do, but in a special way discovered by Gustav Kirchhoff (1824–1887) in 1859. In particular, Kirchhoff argued that the ratio e/a is independent of the material of the absorber and emitter when that material is in thermal equilibrium with the radiation it absorbs and emits.

For a proof of Kirchhoff's law, consider a material body at temperature T surrounded by and in thermal equilibrium with radiation of energy density $u(T)$. Radiation is incident upon a unit area of the surface of this material at a rate proportional to the product $c \cdot u(T)$ of the speed of light c and the radiant energy density $u(T)$. Therefore, a unit area of the surface *absorbs* energy at a rate proportional to $a \cdot c \cdot u(T)$, reflects energy at a rate proportional to $(1-a) \cdot c \cdot u(T)$, and emits energy per unit area at a rate equal to its emissive power e. In equilibrium the emissive and absorption rates must equal one another, and so

$$e \propto a \cdot c \cdot u(T). \tag{3.2}$$

The proportionality constant that turns (3.2) into an equality is, as is shown elsewhere,[1] the geometrical factor 1/4. Therefore,

$$\frac{e}{a} = \frac{c \cdot u(T)}{4}, \tag{3.3}$$

where the right-hand side depends only on universal constants and the radiation temperature T but not at all on the composition of the absorbing and emitting material. Therefore, the emissive power e of a material divided by its coefficient of absorption a is a function only of the temperature T of the radiation with which the material is in equilibrium and is independent of the emitting and absorbing material—a statement that comprises part of *Kirchhoff's law*, the part that formalizes Prevost's and Rumford's observation that a good absorber of radiant energy is a good emitter and a bad absorber of radiant energy is a bad emitter.

A complete statement of Kirchhoff's law is that the ratio of the spectral emissive power e_λ to the spectral coefficient of absorption a_λ, each applying to wavelengths between λ and $\lambda + d\lambda$, is such that

$$\frac{e_\lambda}{a_\lambda} = \frac{c \cdot u_\lambda(T,\lambda)}{4} \tag{3.4}$$

where $u_\lambda(T,\lambda)$ is the *spectral energy density of blackbody radiation*. The proof of this statement is identical to that of the previous statement (3.3) except that we place a filter in front of the unit area of surface so that it only transmits radiation with wavelengths between λ and $\lambda + d\lambda$.

An important consequence of Kirchhoff's spectral law (3.4) is that when the emitting and absorbing body is a *blackbody*, so-called because the body absorbs all radiation that falls upon it and so $a_\lambda = 1$ for all wavelengths λ, the spectral emissive power e_λ is, according to (3.4), given by

$$e_\lambda(T,\lambda) = \frac{c \cdot u_\lambda(T,\lambda)}{4}. \tag{3.5}$$

However, as a blackbody radiation experimentalist quickly learns, only the spectral emissive power e_λ of a particular object can be directly measured. Therefore, experimentalists take care to determine or account for the spectral coefficient of absorption a_λ of that object and to use (3.4) rather than (3.5) to determine a value for spectral energy density $u_\lambda(T,\lambda)$.

Planck found attractive the idea of discovering the universal function $u_\lambda(T,\lambda)$, which we call the spectral energy density of blackbody radiation. For, as he later said, its structure is "independent of special bodies and substances," which will "necessarily retain its importance for all times and cultures, even for non-terrestrial and non-human ones."[2]

4 The Stefan-Boltzmann Law, 1884

4.1 Radiation Pressure

Electromagnetic radiation transports energy from one place to another. And, of course, directed energy carries momentum. Indeed, the relation between the energy E in a bundle of unidirectional electromagnetic waves and its momentum p,

$$E = cp, \tag{4.1}$$

follows from electromagnetic theory as worked out by James Clerk Maxwell in 1874.[1] Starting from this result we will derive, as did Boltzmann in 1884, the relationship between the energy density u of blackbody radiation and the isotropic pressure P it exerts.

First, consider the special case of a right, cylindrically shaped bundle of unidirectional, electromagnetic radiation with energy density u that approaches a flat boundary with the direction of its motion parallel to its cylindrical axis and normal to the boundary—as illustrated in Figure 4.1. If the volume of the cylinder is AL where A is its cross-sectional area and L is its length, the electromagnetic energy $E[= uAL]$ and the momentum E/c it contains produces an average force F at the boundary (at which it is absorbed), during the interval L/c, given by

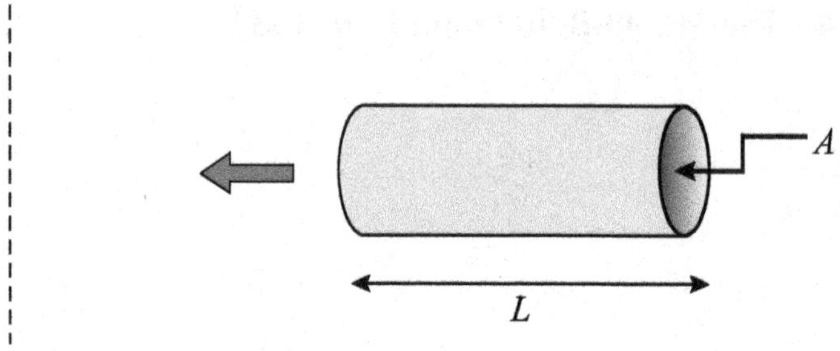

Figure 4.1.
A cylinder of unidirectional, electromagnetic energy approaching a flat boundary in a direction normal to that boundary.

$$\begin{aligned} F &= \frac{E/c}{L/c} \\ &= \frac{E}{L}. \end{aligned} \tag{4.2}$$

Consequently, the pressure $P[=F/A]$ exerted by the radiation at the boundary is given by

$$\begin{aligned} P &= \frac{E}{AL} \\ &= u. \end{aligned} \tag{4.3}$$

Therefore, the pressure P exerted on a surface by unidirectional, electromagnetic waves, normal to that surface, is equal to their energy density u.

When the radiation is isotropic rather than unidirectional, the situation is different. In this case, most of the radiation falls upon the boundary at angles oblique to its normal as shown in Figure 4.2. Therefore, the intensity of the incident energy per unit area is reduced below u by a factor of $\cos\theta$ where θ is the angle the axis of the cylinder of radiation makes with the normal to the boundary. And the momentum delivered

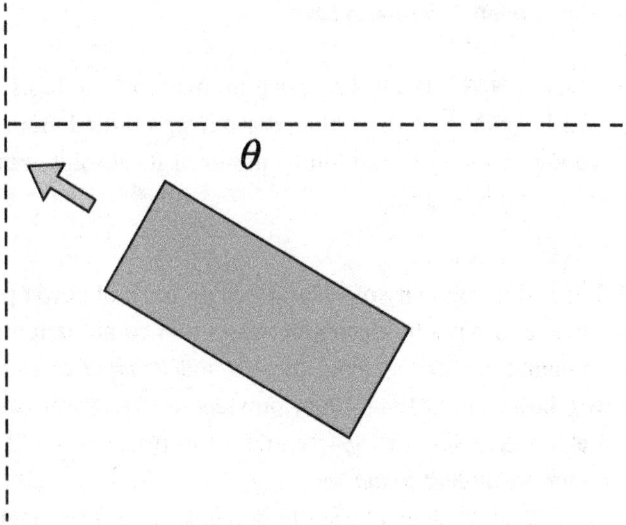

Figure 4.2.
A cylinder of electromagnetic radiation approaching a flat boundary at an angle θ normal to the boundary.

normal to the boundary is also reduced by a factor of $\cos\theta$. For these reasons, the pressure P exerted by isotropic radiation falling on the surface averaged over all angles of incidence 0 to $\pi/2$ is given by

$$P = u\frac{\int_0^{\pi/2}\cos^2\theta\sin\theta d\theta}{\int_0^{\pi/2}\sin\theta d\theta}$$

$$= \frac{u}{3}. \tag{4.4}$$

Equation (4.4) establishes the equation of state $P = u/3$ that relates the pressure P and the energy density u of an electromagnetic fluid of isotropic radiation.

4.2 The Stefan-Boltzmann Law

Josef Stefan (1835–1893) discerned, in 1879, on the basis of previously published data, that the energy emitted from a blackbody increases as the fourth power of its absolute temperature, that is, so that

$$e = \sigma T^4 \tag{4.5}$$

where e is the emissivity of a blackbody (in units of power per unit area) and σ is a fundamental constant or combination of fundamental constants called the *Stefan-Boltzmann constant.* Ludwig Boltzmann (1844–1906) provided a theoretical basis for Stefan's discovery in 1884,[2] and, for this reason, $e = \sigma T^4$ is called the *Stefan-Boltzmann law.*

Our derivation depends, as did Boltzmann's, on two facts. First, we know from Maxwell's electrodynamics and the geometry of isotropic radiation that the pressure P exerted by blackbody radiation is related to its energy density u by

$$P = \frac{u}{3}. \tag{4.6}$$

Secondly, we know, from section 3.2, that the energy density u of blackbody radiation at a given temperature T is independent of its volume V so that

$$\left(\frac{\partial u}{\partial V}\right)_T = 0. \tag{4.7}$$

Equations (4.5), (4.6), and (4.7) summarize what Boltzmann knew of blackbody radiation in 1884.

Boltzmann quite boldly applied the laws of thermodynamics to a system composed of nothing but blackbody radiation. In doing so he concentrated on determining the form of the function $u(T)$ since the emissivity e of a blackbody is related, as given by (3.3), to the energy density u of blackbody radiation by

$$e = \frac{c}{4}u \tag{4.8}$$

when the absorption coefficient $a = 1$.

Our derivation in this section diverges from the pattern of Boltzmann's, which is reproduced in section 4.7, and also from the alternative derivation of section 4.4 while following a logic that has become common at least since 1912 when Planck published the second edition of his *Theory of Heat Radiation*.[3]

Imagine a system of isotropic radiation with total volume V. The first and second laws of thermodynamics relate changes among the state variables of this system including its entropy S and its temperature T via the fundamental constraint

$$d(uV) = TdS - PdV. \tag{4.9}$$

Any change in these quantities not satisfying (4.9) violates either one or both of the laws of classical thermodynamics.

Given the equation of state $P = u/3$, the fundamental constraint (4.9) becomes

$$d(uV) = TdS - \frac{u}{3}dV. \tag{4.10}$$

Rewriting (4.10) as

$$dS = \frac{4u}{3T}dV + \frac{V}{T}du, \tag{4.11}$$

we note that the entropy is a function of two variables, u and V, so that $S = S(u,V)$. Here we follow the formalism established in section 2.4, by equating the cross derivatives of the formal equations of state, $(\partial S/\partial V)_u = 4u/3T$ and $(\partial S/\partial u)_V = V/T$. In this way we find that

$$\frac{\partial}{\partial u}\left(\frac{4u}{3T}\right)_V = \frac{\partial}{\partial V}\left(\frac{V}{T}\right)_u. \tag{4.12}$$

To summarize, requirement (4.12) follows from the equation of state $P = u/3$ and the first and second laws of thermodynamics as encapsulated in the fundamental constraint (4.9).

Expanding the partial derivatives in (4.12) we find that

$$\frac{4}{3T} - \frac{4u}{3T^2}\left(\frac{\partial T}{\partial u}\right)_V = \frac{1}{T}, \tag{4.13}$$

which is equivalent to

$$\frac{1}{3T} = \frac{4u}{3T^2}\frac{dT}{du} \tag{4.14}$$

where, since u is not a function of V, we have replaced the partial derivative $(\partial T/\partial u)_V$ with the total derivative dT/du. Separating variables we find that

$$\frac{du}{u} = 4\frac{dT}{T}, \tag{4.15}$$

and so

$$u = aT^4 \tag{4.16}$$

where a is called the *radiation constant.* While the value of the radiation constant a is not determined by this derivation, it can be measured, and is related to the Stefan-Boltzmann constant σ, through (4.5), (4.8), and (4.16), by

$$a = \frac{4\sigma}{c}. \tag{4.17}$$

4.3 The Adiabatic Invariant of Blackbody Radiation

Recall that the adiabatic invariant of a system is any combination of its thermodynamic variables that remains constant during an isentropic process and that an isentropic, or entropy-conserving, process is necessarily reversible and one in which the system does not receive or give up heat. Furthermore, a reversible process is one that is both quasistatic and without friction or internal dissipation.

The adiabatic invariant of blackbody radiation is easily determined from the equation of state $P = u/3$ and the fundamental constraint for a fluid with $dS = 0$, that is, from

$$d(uV) = -\frac{u}{3}dV, \tag{4.18}$$

which is equivalent to

$$\frac{du}{u} + \frac{4}{3}\frac{dV}{V} = 0. \tag{4.19}$$

The latter is easily integrated to give

$$uV^{4/3} = ad.inv.. \tag{4.20}$$

Other forms of the adiabatic invariant of blackbody radiation, for instance $PV^{4/3}$ and $TV^{1/3}$, can be generated by replacing u in (4.20) with either P or T via the equations of state $P = u/3$ and $u = aT^4$.

4.4 An Alternate Derivation of the Stefan-Boltzmann Law

The fundamental constraint for the system of blackbody radiation (4.11) may also be written in the form

$$\begin{aligned} dS &= \left(\frac{uV}{T}\right)\left[\frac{4dV}{3V} + \frac{du}{u}\right] \\ &= \left(\frac{uV}{T}\right) d\left[\ln\left(uV^{4/3}\right)\right], \end{aligned} \tag{4.21}$$

according to which the adiabatic invariant of blackbody radiation $uV^{4/3}$ changes if and only if the entropy S of blackbody radiation changes. This means that

$$S(u,V) = S(uV^{4/3}). \tag{4.22}$$

Consequently, the number of variables upon which the entropy depends is reduced from two to one. This reduction is key to the following "alternate derivation" of the Stefan-Boltzmann law.

Comparing the fundamental constraint for this system in the form

$$dS = \frac{4u}{3T}dV + \frac{V}{T}du \tag{4.23}$$

with the identity

$$dS = \left(\frac{\partial S}{\partial V}\right)_u dV + \left(\frac{\partial S}{\partial u}\right)_V du \tag{4.24}$$

produces the following (formal) equations of state

$$\left(\frac{\partial S}{\partial V}\right)_u = \frac{4u}{3T} \tag{4.25}$$

and

$$\left(\frac{\partial S}{\partial u}\right)_V = \frac{V}{T}. \tag{4.26}$$

Using the chain rule to differentiate the single-variable entropy dependence (4.22) leads to

$$\left(\frac{\partial S}{\partial u}\right)_V = S'(uV^{4/3})V^{4/3} \tag{4.27}$$

where the notation $S'(uV^{4/3})$ means differentiation with respect to the argument $uV^{4/3}$. When combined with (4.26), relation (4.27) leads to $S'(uV^{4/3})V^{4/3} = V/T$ or, equivalently, to

$$S'(uV^{4/3}) = \frac{1}{(V^{1/3}T)}. \tag{4.28}$$

Since a function of an adiabatic invariant is also an adiabatic invariant, equation (4.28) demonstrates that $V^{1/3}T$ is another version of the adiabatic invariant of blackbody radiation.

Given that $uV^{4/3}$ and $V^{1/3}T$ are different forms of the adiabatic invariant of blackbody radiation, their relation must express at least part of the content of the equations of state of blackbody radiation. For instance, the two sides of (4.28) produce

$$uV^{4/3} = f\left(V^{1/3}T\right) \tag{4.29}$$

where $f\left(V^{1/3}T\right)$ is an unknown function of a single variable related to the inverse of the function $S'\left(V^{1/3}T\right)$. According to (4.29), the only such function $f\left(V^{1/3}T\right)$ that renders the energy density u a function $u(T)$ of the thermodynamic temperature T alone is that defined by

$$f\left(V^{1/3}T\right) = const.\cdot\left(V^{1/3}T\right)^4. \tag{4.30}$$

Therefore, the functional definition (4.30) and the relation between two adiabatic invariants (4.29) become

$$uV^{4/3} = const.\cdot\left(V^{1/3}T\right)^4, \tag{4.31}$$

from which the Stefan-Boltzmann law

$$u = aT^4 \tag{4.32}$$

follows. Starting this calculation by differentiating the reduced entropy $S\left(uV^{4/3}\right)$ with respect to V and using the other formal equation of state (4.25) leads to the same result.

This alternate derivation of the Stefan-Boltzmann law is no more compact or transparent than the traditional one given in section 4.2. We include it here because it demonstrates a claim made in section 2.9 that *every function relating two different forms of the adiabatic invariant of a system reveals something of the content of the system's equations of state*, and because we will use this claim to derive Wien's displacement law in chapter 5.

4.5 The Entropy of Blackbody Radiation

Since we now know two independent equations of state of blackbody radiation, $u = aT^4$ and $P = u/3$, we may use them to derive an expression for its entropy. And that entropy, as indicated in (4.22), should be a function of only the adiabatic invariant $uV^{4/3}$ or its equivalent and fundamental or system constants.

Recall, from equation (4.23), that the fundamental constraint for blackbody radiation may be cast in the form

$$dS = \frac{4u}{3T}dV + \frac{V}{T}du. \tag{4.33}$$

Eliminating the absolute temperature T from (4.33) via the equation of state $u = aT^4$ yields

$$dS = \frac{4a^{1/4}u^{3/4}}{3}dV + \frac{a^{1/4}V}{u^{1/4}}du. \tag{4.34}$$

Therefore, given the identity $dS = (\partial S/\partial V)_u dV + (\partial S/\partial u)_V du$,

$$\left(\frac{\partial S}{\partial V}\right)_u = \frac{4a^{1/4}u^{3/4}}{3} \tag{4.35}$$

and

$$\left(\frac{\partial S}{\partial u}\right)_V = \frac{a^{1/4}V}{u^{1/4}}. \tag{4.36}$$

Integrating these equations, (4.35) and (4.36), produces, respectively,

$$S(u,V) = \frac{4a^{1/4}u^{3/4}V}{3} + f(u) \tag{4.37}$$

and

$$S(u,V) = \frac{4a^{1/4}u^{3/4}V}{3} + g(V) \tag{4.38}$$

where the functions $f(u)$ and $g(V)$ are as yet undetermined.

The only way these two expressions, (4.37) and (4.38), for the entropy $S(u,V)$ can be made consistent with one another is for $f(u) = g(V) = const.$ so that

$$S(u,V) = \frac{4a^{1/4}u^{3/4}V}{3} + const. \tag{4.39}$$

where *const.* stands for a function of fundamental or system constants. But there are no fundamental or system constants of blackbody radiation apart from the radiation constant a (or, equivalently, the Stefan-Boltzmann constant σ). Furthermore, in Boltzmann's time there were no other fundamental constants that, when combined with a or σ, could produce a quantity with the dimensions of entropy. When the energy density of blackbody radiation u or its volume V vanishes, it is reasonable to assume that its entropy also vanishes. For this reason (4.39) is usually written as

$$S(u,V) = \frac{4a^{1/4}u^{3/4}V}{3}. \tag{4.40}$$

4.6 The Universality of Blackbody Radiation

There are a number of similarities between our analyses of an ideal gas and of blackbody radiation. Each is an isotropic fluid with two independent variables and two independent equations of state. These thermodynamic variables are, in each case, those of an isotropic fluid: internal energy U or energy density $u[= U/V]$, thermodynamic temperature T, entropy S, pressure P, and volume V. And, from their respective equations of state, an entropy function can be derived.

But these similarities mask profound differences. The ideal gas equations of state are only approximations that, in the late nineteenth century, were known to break down at low temperatures and high densities.[4] Furthermore, the equations of state of an ideal gas contain parameters or system constants, such as mole number n and heat capacity C_V, that characterize the particular gas in question. The molar specific heat C_V/n, in particular, is different for gases composed of different kinds of molecules in different temperature regimes.

On the other hand, the equations of state for blackbody radiation, $P = u/3$ and $u = aT^4$, have no such system parameters. The constant a is a fundamental constant or a combination of fundamental constants. Unlike the ideal gas equations of state, those of blackbody radiation accurately apply in all temperature, pressure, and energy density regimes.

It was this accuracy and this universality that attracted Boltzmann, Wien, Planck, and others to blackbody radiation. If one could discover an explanation of blackbody radiation in terms of foundational principles that underlie its quite successful thermodynamic description, one would discover something fundamental about the world.

4.7 Boltzmann's 1884 Derivation

Appendix A contains Louis Buchholtz's translation of Boltzmann's 1884 paper, "A Derivation of Stefan's Law, Concerning the Temperature Dependence of Thermal Radiation, from the Electromagnetic Theory of Light." Modern readers will find that Boltzmann's paper does not conform to current standards of clarity. Even so, the paper reveals what Boltzmann did and did not take for granted and as such opens a window into his mental world.

For instance, Boltzmann did not begin his description of blackbody radiation with the fundamental constraint of an isotropic fluid. Nor did he mention the word *entropy*, nor that the process Boltzmann describes is a Carnot cycle. Even so, the fundamental constraint, the concept of entropy, and that the fluid returns to its starting point are crucial to his derivation.

Instead, Boltzmann imagined a two-chambered cylinder with a movable piston head separating the radiation it contains into two thermally isolated systems. Therefore, when the piston expands one chamber, it necessarily shrinks the other. Yet Boltzmann did not help his readers with drawings that illustrate these relations. In the following, we supply the missing drawings and the missing intellectual signposts. We also modernize Boltzmann's notation while faithfully reproducing the logic of his derivation.

Note that Boltzmann knew that the energy density u of blackbody radiation is a function $u(T)$ of the thermodynamic temperature T of the radiation alone and that the pressure P exerted by radiation is related to the energy density u by the equation of state $P = u/3$. Even so he carried the quantities u and P (which he denoted, respectively, Ψ and f) as separate variables until the very end of his derivation.

Consider then a long hollow cylinder of cross-sectional area A whose walls prohibit heat flow. The cylinder is capped at its left end with a thermal reservoir at temperature T_0 and at its right end with a different thermal reservoir at temperature T where $T < T_0$. Inside the cylinder and initially flush with its left end is a movable piston head composed of rigid, adiabatic material. Figure 4.3 illustrates this device and the stages of its procession.

In the first transition, from the situation of Figure 4.3(a) to that of Figure 4.3(b), the piston moves reversibly to the right through distance a. During this movement a chamber is created on the left side of the piston that fills with blackbody

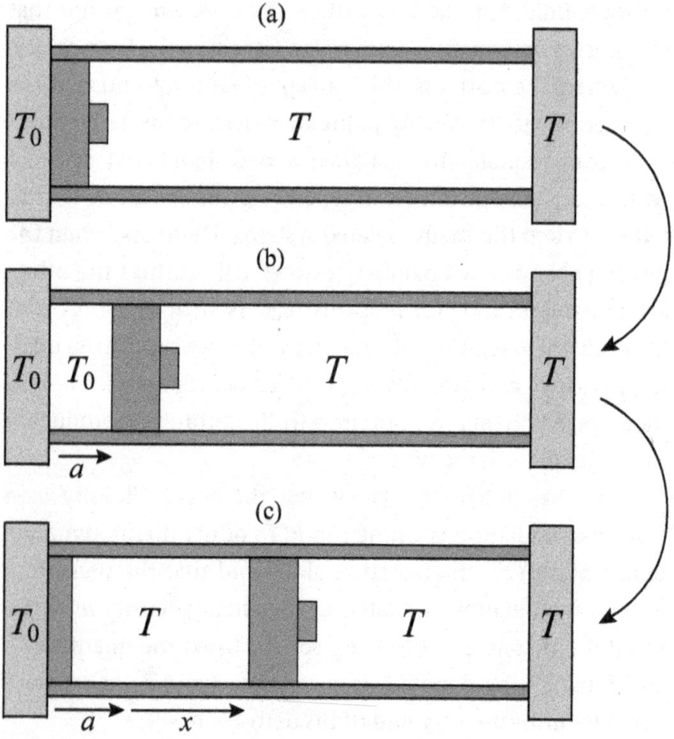

Figure 4.3.
Processes in Boltzmann's derivation. The cylinder is surrounded by rigid, adiabatic material except at its ends, which in parts (a) and (b) are exposed to thermal reservoirs at given temperatures. In part (c) the thermal reservoir at the left end is closed off with an adiabatic barrier.

radiation at temperature T_0. At the same time, as the right chamber shrinks in size, the radiation it contains remains at temperature T.

We use the fundamental constraint $d(uV) = TdS - PdV$ to find the entropy increase ΔS_L of the radiation on the left side of the piston caused by this transition. In this way, we find that $u\Delta V = T\Delta S_L - P\Delta V$ and so

$$\Delta S_L = \frac{[u(T_0)+P(T_0)]}{T_0}\Delta V_a \tag{4.41}$$

where $\Delta V_a = aA$. At the same time, the entropy of the radiation on the right side of the piston decreases by ΔS_R where

$$\Delta S_R = \frac{-[u(T)+P(T)]}{T}\Delta V_a \tag{4.42}$$

and where, again, $\Delta V_a = aA$. Therefore, during the first process the total change $\Delta S_1 [= \Delta S_L + \Delta S_R]$ in the entropy of the system, composed of the radiation in both chambers, is given by

$$\Delta S_1 = \frac{[u(T_0)+P(T_0)]\Delta V_a}{T_0} - \frac{[u(T)+P(T)]\Delta V_a}{T}. \tag{4.43}$$

Preliminary to the second transition from the situation of Figure 4.3(b) to that of Figure 4.3(c), an adiabatic barrier, isolating the T_0 reservoir, is placed at the left end of the cylinder. Then the piston moves reversibly to the right a further distance x, where x is defined to be as far as necessary to adiabatically depress the temperature of the radiation on the left side of the piston to match the temperature T isothermally maintained on the right side of the piston. During this second process the entropy of the radiation on the left side of the piston is unchanged, because it expands reversibly and adiabatically, while the entropy of the radiation on the right side of the piston decreases by an amount ΔS_2 given by

$$\Delta S_2 = \frac{-[u(T)+P(T)]}{T}\Delta V_x \tag{4.44}$$

where $\Delta V_x = xA$.

Since the final state of the blackbody radiation is the same as its initial state, the net entropy change of the system composed of both chambers on either side of the piston vanishes. Since both processes are reversible, $\Delta S_1 + \Delta S_2 = 0$ or, equivalently,

$$\frac{[u(T_o)+P(T_o)]\Delta V_a}{T_o} - \frac{[u(T)+P(T)]\Delta V_{a+x}}{T} = 0, \tag{4.45}$$

which is equivalent to

$$\frac{[u(T_o)+P(T_o)]\Delta V_a}{T_o} = \frac{[u(T)+P(T)]\Delta V_{a+x}}{T} \tag{4.46}$$

where $\Delta V_{a+x} = (a+x)A$.

At this point Boltzmann imagined that the temperature T of the reservoir at the right end of the cylinder is variable while the *definitions* of the other quantities, in particular the distance x, are maintained. For instance, if the temperature T increases, the distance x through which the piston must travel, in order to adiabatically depress the temperature of the radiation on the left until it matches the temperature T of the radiation on the right, decreases. For this reason, Boltzmann was justified in considering the quantities T and ΔV_{a+x}, or, equivalently, $V[=\Delta V_{a+x} = (a+x)A]$, to be related and variable while the quantities ΔV_a and T_o remain constant.

In this case, the right-hand side of (4.46), namely $V[u+P]/T$, remains constant, while the quantities composing it, namely V, u, P, and T, vary. In other words, $V[u+P]/T$, or equivalently Vu/T, is an adiabatic invariant. Therefore, during the second process in which the blackbody radiation on the left side of the piston is adiabatically expanded, the equations

$$d\left[\frac{V(u+P)}{T}\right]=0 \tag{4.47}$$

and

$$d[uV]=-PdV \tag{4.48}$$

link changes in the variables $u(T)$, $P(T)$, T, and V. Equation (4.48) is, of course, the fundamental constraint with $dS = 0$. For these reasons, equations (4.47) and (4.48) describe an adiabatic or isentropic process.

From (4.47) we have

$$\frac{d[V(u+P)]}{T}=\frac{[V(u+P)]}{T^2}dT \tag{4.49}$$

which, upon expansion, becomes

$$d(Vu)+PdV+VdP=\frac{V}{T}(u+P)dT. \tag{4.50}$$

Equation (4.48) reduces (4.50) to

$$TdP=(u+P)dT, \tag{4.51}$$

which reproduces the last equation of Boltzmann's 1884 paper.

Only a little algebra separates (4.51) from something that *looks like* the Stefan-Boltzmann law. Boltzmann outlines this algebra in the second paragraph of his paper. In particular, since $P = u/3$, we eliminate P from equation (4.51) and, in this way, produce

$$Tdu=4udT. \tag{4.52}$$

Separating variables and integrating both sides of (4.52) yields

$$\frac{u(T)}{T^4}=ad.inv.. \tag{4.53}$$

We use the symbol *ad.inv.* on the right side of (4.53) because the equations from which this result derives, namely (4.47) and (4.48), obtain only during an isentropic process.

It is in this way that Boltzmann discovered, not an equation of state but rather an adiabatic invariant of blackbody radiation. While he recognized "the decidedly provisional character of the derivation," Boltzmann says nothing that would transform the combination u/T^4 from an adiabatic to an absolute invariant, that is, to a fundamental constant or a combination of fundamental constants. For instance, Boltzmann could have supposed that $u = aT^4$ and $P = u/3$ are independent equations of state and shown that these were consistent with the two laws of classical thermodynamics as expressed in the fundamental constraint. However, he does not take this step.

In spite of its shortcomings, Boltzmann's 1884 paper is a skillful probing of the physics of blackbody radiation. And, if it does not contain a convincing derivation of what today is known as the Stefan-Boltzmann law, the paper did pave the way for a deeper understanding of the thermodynamics of blackbody radiation and for the more convincing derivations of the Stefan-Boltzmann law soon to follow.[5]

5 Wien's Contributions, 1893–1896

5.1 Spectral Energy Density

The universality of blackbody radiation attracted Max Planck and others to the problem of explaining its properties. Planck, in particular, sought to derive the spectral energy density function $u_\lambda(T,\lambda)$ of blackbody radiation from fundamental principles, that is, to derive the blackbody energy per unit volume at temperature T in an interval of wavelengths between λ and $\lambda + d\lambda$.

The spectral energy density u_λ is related to the total energy density u by

$$u(T) = \int_0^\infty u_\lambda(T,\lambda) d\lambda. \tag{5.1}$$

Therefore, given the Stefan-Boltzmann law $u = aT^4$, the integration

$$aT^4 = \int_0^\infty u_\lambda(T,\lambda) d\lambda \tag{5.2}$$

is required of the function $u_\lambda(T,\lambda)$ where $a\,[= 4\sigma/c]$ is the radiation constant and σ the Stefan-Boltzmann constant.

In 1893 Wilhelm Wien took a significant step toward deriving the function $u_\lambda(T,\lambda)$ by demonstrating that the spectral energy density must assume the form

$$u_\lambda(T,\lambda) = T^5 h(T\lambda) \tag{5.3}$$

where $h(T\lambda)$ is an undetermined function[1]—a result known as *Wien's displacement law.*[2] In deriving this result Wien employed the constraints of classical thermodynamics and the identity of an "electromagnetic adiabatic invariant." However, Wien's derivation of his displacement law and variations of this derivation by Planck,[3] Buckingham,[4] Born,[5] and others[6] seem difficult and opaque.

The hallmark of our alternative derivation is its manifest thermodynamic character and its reliance on adiabatic invariants as modeled in the alternate derivation of the Stefan-Boltzmann law in section 4.4. Recall that we use the term *adiabatic invariant* to refer to quantities that are invariant during an isentropic or entropy-conserving process.

5.2 Cumulative Spectral Energy Density

The thermodynamic system we adopt for the purpose of deriving Wien's displacement law is that composed of isotropic blackbody radiation within the interval of wavelengths from 0 to λ. Boltzmann,[7] Planck,[8] and Einstein[9] all realized that a subset of blackbody radiation composed of wavelengths within a certain range of wavelengths or frequencies constitutes a thermodynamic system characterized by a thermodynamic temperature T.

The energy of the system we consider is denoted $VU_0^\lambda(T)$ where $U_0^\lambda(T)$ is an energy density, which we call the *cumulative spectral energy density*, defined by

$$U_0^\lambda(T) = \int_0^\lambda u_x(T,x)dx. \tag{5.4}$$

Here the sub- and superscripts on the right-hand side of U_0^λ refer to the integration limits in (5.4). Therefore, U_0^λ is an energy density of a system composed of blackbody radiation with wavelengths between 0 and λ, where T is its temperature. The limiting value of the cumulative spectral energy density is given by $U_0^\infty(T) = u(T) = aT^4$.

The notation used in (5.4) should help us distinguish the cumulative spectral energy density U_0^λ from similar symbols introduced here and in chapters 2, 3, and 4: U (the energy of a system), u (the energy density), and u_λ (the spectral energy density of blackbody radiation). Note that U_0^λ is short for $U_0^\lambda(T)$ and that, given definition (5.4), the cumulative spectral energy density is an explicit function of both T and λ but not of the system volume V. Also following from (5.4) is the relation

$$\left(\frac{\partial U_0^\lambda}{\partial \lambda}\right)_T = u_\lambda(T,\lambda) \tag{5.5}$$

between the partial derivative $(\partial U_0^\lambda/\partial\lambda)_T$ of the cumulative spectral energy density U_0^λ with respect to λ and the spectral energy density $u_\lambda(T,\lambda)$. This relation will enable us to "look under the integral sign" of definition (5.4).

5.3 Thermodynamic Adiabatic Invariants

The fundamental constraint for the system composed of isotropic blackbody radiation with wavelengths between 0 and λ is

$$d(VU_0^\lambda) = TdS_0^\lambda - P_0^\lambda dV \tag{5.6}$$

where the volume V, the cumulative spectral energy density U_0^λ, the temperature T, the entropy S_0^λ, and the pressure P_0^λ are the fluid variables that describe this system. We know one of the system's two independent equations of state, that is, we know that

$$P_0^\lambda = \frac{U_0^\lambda}{3}, \tag{5.7}$$

a relation that holds for all cases of isotropic radiation. Given (5.7), the fundamental constraint (5.6) becomes

$$d\left(VU_0^\lambda\right) = TdS_0^\lambda - \frac{U_0^\lambda}{3} dV, \tag{5.8}$$

which may be rewritten as

$$\begin{aligned} TdS_0^\lambda &= d\left(VU_0^\lambda\right) + \frac{U_0^\lambda}{3} dV \\ &= \frac{4U_0^\lambda}{3} dV + VdU_0^\lambda \\ &= \left(VU_0^\lambda\right)\left[\frac{4dV}{3V} + \frac{dU_0^\lambda}{U_0^\lambda}\right] \\ &= \left(VU_0^\lambda\right)\left[d\ln\left(V^{4/3}U_0^\lambda\right)\right]. \end{aligned} \tag{5.9}$$

Therefore $V^{4/3}U_0^\lambda$ is an adiabatic invariant of the system, and this adiabatic invariant changes if and only if the entropy S_0^λ of the system changes.

Thus, from (5.9), we find that

$$V^{4/3}U_0^\lambda = ad.inv. \tag{5.10}$$

where *ad.inv.* stands for *adiabatic invariant.* We use the notation *ad.inv.*, as we did in chapters 2 and 4, in order to emphasize that an adiabatic invariant is not the same as an absolute invariant. Rather, an adiabatic invariant is invariant only during an isentropic process.

Therefore, according to (5.9), it must be that

$$S_0^\lambda\left(U_0^\lambda, V\right) = S_0^\lambda\left(V^{4/3}U_0^\lambda\right) \tag{5.11}$$

where the left-hand side of (5.11) is a function of two variables, U_0^λ and V, and the right-hand side is necessarily a different function of one variable $V^{4/3}U_0^\lambda$. In this way, knowing the adiabatic invariant allows us to reduce, from two to one, the number of variables upon which the entropy of blackbody radiation depends.

We exploit (5.11), as we exploited an analogous expression in section 4.4, by noting that the fundamental constraint (5.8) may be expressed as

$$dS_0^\lambda = \frac{4U_0^\lambda}{3T}dV + \frac{V}{T}dU_0^\lambda, \tag{5.12}$$

which, given the identity

$$dS_0^\lambda = \left(\frac{\partial S_0^\lambda}{\partial V}\right)_{U_0^\lambda} dV + \left(\frac{\partial S_0^\lambda}{\partial U_0^\lambda}\right)_V dU_0^\lambda, \tag{5.13}$$

produces two (formal) equations of state:

$$\left(\frac{\partial S_0^\lambda}{\partial V}\right)_{U_0^\lambda} = \frac{4U_0^\lambda}{3T} \tag{5.14}$$

and

$$\left(\frac{\partial S_0^\lambda}{\partial U_0^\lambda}\right)_V = \frac{V}{T}. \tag{5.15}$$

The derivative of the entropy dependence $S_0^\lambda\left(V^{4/3}U_0^\lambda\right)$ with respect to U_0^λ and the formal equation of state (5.15) together produce

$$S_0^{\lambda\prime}\left(V^{4/3}U_0^\lambda\right)V^{4/3} = \frac{V}{T}, \tag{5.16}$$

or, equivalently,

$$S_0^{\lambda\prime}\left(V^{4/3}U_0^{\lambda}\right)=\frac{1}{V^{1/3}T}. \tag{5.17}$$

Equation (5.17) also follows from the derivative of $S_0^{\lambda}\left(V^{4/3}U_0^{\lambda}\right)$ with respect to V and the corresponding formal equation of state (5.14).

Since a function of an adiabatic invariant is an adiabatic invariant, we conclude from (5.17) that $V^{1/3}T$ is a form of the adiabatic invariant of blackbody radiation, that is,

$$V^{1/3}T = ad.inv.. \tag{5.18}$$

Equations (5.10) and (5.18) identify two forms of the thermodynamic adiabatic invariant of the system, $V^{4/3}U_0^{\lambda}$ and $V^{1/3}T$, that are crucial components of our derivation of the Wien displacement law. A third adiabatic invariant is identified in the next section.

5.4 Wien's Electromagnetic Adiabatic Invariant

The usual way of deriving Wien's electromagnetic invariant is to imagine a short pulse of monochromatic light of wavelength λ that bounces around inside a closed container with reflecting interior surfaces and volume V, while that container slowly and self-similarly shrinks in size.[10] During this process the surfaces of the shrinking container Doppler-shift the reflecting pulse of radiation in such a way that the quantity $V^{1/3}\lambda^{-1}$ remains invariant—adiabatically invariant, as we shall soon argue. Since the thermodynamics of an isotropic fluid is independent of container shape, one may safely assume a shape convenient for this kind of calculation. We outline this approach in Appendix C.

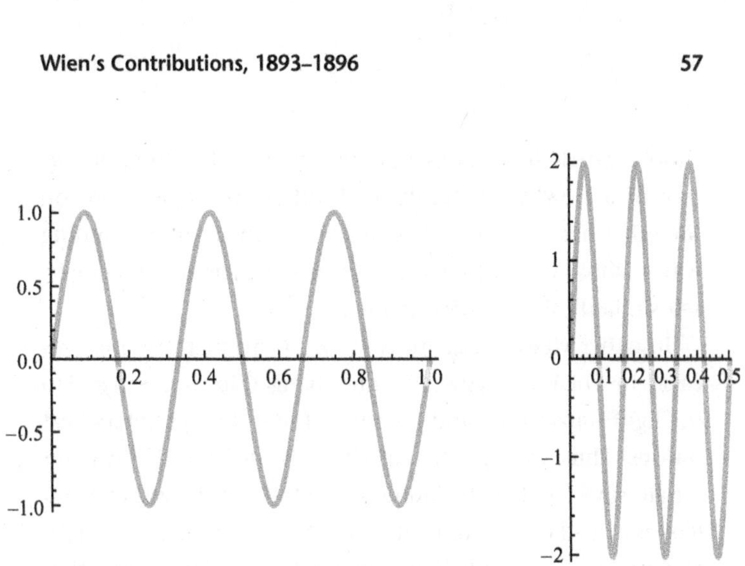

Figure 5.1.
Snapshots of a shrinking one-dimensional container in which definite boundary conditions on the wave are maintained. As the size of this box shrinks from 1 unit (on the left) to 1/2 unit (on the right), the wavelength shrinks by a factor of 2. The amplitude also changes.

But what if the wavelength of the monochromatic radiation is long compared to the linear dimension $V^{1/3}$ of the reflecting, self-similarly shrinking container? Then one would expect the monochromatic radiation to set up standing waves in the container that observe definite boundary conditions at its reflecting surfaces. As the container slowly and self-similarly shrinks, these boundary conditions are maintained and the wavelengths λ of the standing waves decrease in direct proportion to $V^{1/3}$—as illustrated in the one-dimensional case of Figure 5.1. Therefore, again

$$V^{1/3}\lambda^{-1} = ad.inv.. \tag{5.19}$$

It is remarkable that two very different descriptions of electromagnetic radiation produce the same electromagnetic invariant. On one hand, the radiation may be in the form of short, particle-like pulses that reflect from the interior surface of a slowly, self-similarly shrinking container and, on the other hand, the radiation may be in the form of standing waves whose values are fixed at the interior surface of a slowly, self-similarly shrinking container.

In either view, the quantity $V^{1/3}\lambda^{-1}$ is invariant during processes in which the system volume changes slowly compared to the light transit time and for which the system prohibits heat transfer. Thus, we suppose that $V^{1/3}\lambda^{-1}$ is adiabatically invariant for all wavelengths λ including the cutoff wavelength of a system described by the cumulative spectral energy density $U_0^\lambda(T)$.

But is $V^{1/3}\lambda^{-1}$ an adiabatic invariant in the same sense that the thermodynamic quantities $V^{4/3}U_0^\lambda$ and $V^{1/3}T$ are adiabatic invariants? Wien certainly treated the electromagnetic invariant $V^{1/3}\lambda^{-1}$ as if it had the same status as the two thermodynamic adiabatic invariants, $V^{4/3}U_0^\lambda$ and $V^{1/3}T$, for he combined all three in order to create new adiabatic invariants. But was he right to do so?

Paul Ehrenfest was the first to explore the general theory of single-wave or single-particle quantities that are invariant under slow changes of the systems of which they were a part.[11] For instance, the length of a pendulum multiplied by the frequency of its oscillation is an adiabatic invariant for slow changes of the pendulum length. In this way, a long pendulum with low frequency is transformed into a short pendulum with high frequency when the pendulum length is slowly shortened. In 1914 Einstein attached the name "adiabatic hypothesis" to Ehrenfest's principle according to which adiabatic invariants govern the way "allowed motions are transformed into allowed motions."

All these invariants, thermodynamic, electromagnetic, and mechanical, are properly adiabatic invariants for the reason that all are invariant if and only if the process to which they are subject is slow and excludes friction, dissipation, and heat transfer. These conditions are exactly those that define an isentropic process.

5.5 Wien's Displacement Law

All three of the quantities $V^{4/3}U_0^\lambda$, $V^{1/3}T$, and $V^{1/3}\lambda^{-1}$ are adiabatic invariants. The first two are thermodynamic ones (because they are derived from the fundamental thermodynamic constraint) and the third $V^{1/3}\lambda^{-1}$ is an electromagnetic invariant (because it is derived from the properties of electromagnetic waves).

Let's review what we have learned, in sections 2.8 and 4.4, about the relationship among the different forms of the adiabatic invariant of a system. First, any one form of the adiabatic invariant of a system changes during a process if and only if every other form of the adiabatic invariant also changes. From this it follows that every single form of the adiabatic invariant is a function of every other single form. And each of these functional relations expresses part of the content of the equations of state of the system.

We also know that multiplying or dividing two known adiabatic invariants generates other forms of the adiabatic invariant. Therefore, the quotients

$$\frac{V^{4/3}U_0^\lambda}{\left(V^{1/3}T\right)^4} = U_0^\lambda T^{-4} = ad.inv. \tag{5.20}$$

and

$$\frac{V^{1/3}T}{V^{1/3}\lambda^{-1}} = T\lambda = ad.inv. \tag{5.21}$$

identify other adiabatic invariants. Following from (5.20) and (5.21), we conclude that

$$U_0^\lambda(T) = T^4 f(T\lambda), \tag{5.22}$$

where $f(T\lambda)$ is an undetermined function of its argument. Equation (5.22) expresses part of the content of the equations of state of the system of cumulative spectral energy density. Evidently, the $\lambda \to \infty$ limit of (5.22) reduces it to the Stefan-Boltzmann law $U_0^\infty = \sigma T^4$ if $f(T\lambda) \to \sigma$ as $\lambda \to \infty$.

Wien's displacement law is only a short step beyond equation (5.22). Taking the derivative with respect to λ of both sides of (5.22), given (5.5), that is, given $(\partial U_0^\lambda/\partial\lambda)_T = u_\lambda(T,\lambda)$, produces

$$u_\lambda(T,\lambda) = T^5 h(T\lambda) \tag{5.23}$$

where the function $h(T\lambda)[= f'(T\lambda)]$ is undetermined. Equation (5.23) is Wien's displacement law, which itself is part of the content of an equation of state of a system of monochromatic blackbody radiation with wavelengths between λ and $\lambda + d\lambda$.

Wien's displacement law (5.23) may also be written as $u_\lambda(T,\lambda) = \lambda^{-5}\phi(T\lambda)$ where $\phi(T\lambda) = (T\lambda)^5 h(T\lambda)$. Furthermore, since the radiation wavelength λ and frequency ν are related by the speed of light $c[=\nu\lambda]$, we may define the *frequency-related spectral energy density* $u_\nu(T,\nu)$ by requiring that

$$u_\nu(T,\nu)d\nu = u_\lambda(T,\lambda)d\lambda, \tag{5.24}$$

or, more precisely, that

$$u_\nu(T,\nu) = u_\lambda(T,\lambda)|d\lambda/d\nu| \tag{5.25}$$

where the absolute value sign in $|d\lambda/d\nu|$ is necessary in order to account for the definitions $u_\lambda(T,\lambda) > 0$ and $u_\nu(T,\nu) > 0$ and allow for ν to decrease as λ increases. In this way, we find the frequency form of Wien's displacement law to be

$$u_\nu(T,\nu) = \nu^3\Theta(T/\nu), \tag{5.26}$$

or, equivalently,

$$u_\omega(T,\omega) = \omega^3\Phi(T/\omega) \tag{5.27}$$

where $\omega = 2\pi\nu$ and $\Theta(T/\nu)$ and $\Phi(T/\omega)$ are closely related undetermined functions of their arguments. The frequency forms, (5.26) and (5.27), of Wien's displacement law are favored by theorists since the wavelength λ of electromagnetic radiation is modified by transport through different media while the frequency, ν or ω, is not.

To summarize, this derivation of Wien's displacement law is based on the following:

(1) The cumulative spectral energy density U_0^λ, which is a function of only the system temperature T and the cut-off wavelength λ, is related to the spectral energy density $u_\lambda(T,\lambda)$ by $(\partial U_0^\lambda/\partial\lambda)_T = u_\lambda(T,\lambda)$,

(2) The first and second laws of thermodynamics are encapsulated in the fundamental constraint $d(VU_0^\lambda) = TdS_0^\lambda - P_0^\lambda dV$,

(3) The pressure equation of state for the spectral energy density of isotropic radiation with wavelengths between 0 and λ is $P_0^\lambda = U_0^\lambda/3$,

(4) The thermodynamic adiabatic invariants of the system are $V^{4/3}U_0^\lambda$ and $V^{1/3}T$ while the electromagnetic adiabatic invariant is $V^{1/3}\lambda^{-1}$,

(5) Each form of the adiabatic invariant of a thermodynamic system changes if and only if every other form of the adiabatic invariant of the system changes, and

(6) The relation between different forms of the adiabatic invariant expresses part of the content of the equations of state of the system.

Those seeking another application of the method summarized here should consult Appendix D, "An Ideal Gas 'Displacement Law.'"

5.6 A Dimensional Consequence of Wien's Displacement Law

At the time of Wien's derivation of his displacement law in 1893, the only recognized fundamental constants were the speed of light in vacuum c and the gravitational constant G. After all, 1893 was several years before J. J. Thomson's discovery of the electron and his measurement of its charge-to-mass ratio in 1897. And, even then, it took some time for physicists to recognize that the electron charge e and mass m_e were fundamental constants.

It seems likely that the speed of light c enters into the unknown function $h(T\lambda)$ that characterizes the displacement law

$$u_\lambda(T,\lambda) = T^5 h(T\lambda). \tag{5.28}$$

After all, blackbody radiation is an electromagnetic phenomenon. But, surely, the gravitational constant G has nothing to do with blackbody radiation. For this reason, a dimensional problem arises. How can factors of c, T, and $T\lambda$ by themselves combine to produce the dimensions of the spectral energy density function $u_\lambda(T,\lambda)$?

At least two new fundamental constants are needed: one to normalize the product $T\lambda$ and the other to make the right- and left-hand sides of (5.28) dimensionally homogeneous. However, in 1893 Wien seemed not to recognize this need. It was up to Planck in 1899 to become aware of and impressed by this requirement.[12] Eventually two constants, now called

Planck's and Boltzmann's constants, respectively h and k, arose from Planck's analysis of blackbody radiation.

5.7 A Practical Consequence of Wien's Displacement Law

There are other consequences of Wien's displacement law. For in addition to $u_\lambda(T,\lambda) = T^5 h(\lambda T)$ we know that, according to the Stefan-Boltzmann law,

$$\begin{aligned} \sigma T^4 &= T^5 \int_0^\infty h(T\lambda)\,d\lambda \\ &= T^4 \int_0^\infty h(x)\,dx, \end{aligned} \tag{5.29}$$

and, therefore, that

$$\sigma = \int_0^\infty h(x)\,dx. \tag{5.30}$$

Therefore, we know that, if the function $h(x)$ diverges on its domain, that divergence must be integrable.

In fact, the experimental evidence suggested that $h(x)$ has a single, finite, absolute maximum on the positive real axis. For this reason, the spectral energy density of blackbody radiation $u_\lambda(T,\lambda)$ reaches peak intensity as a function of λ at a value of $T\lambda$ that maximizes $h(T\lambda)$. We denote the product $T\lambda$ that maximizes this function by Λ. Therefore, the wavelength λ_m that maximizes the spectral energy density is related to the system temperature T by

$$\lambda_{\text{max}} = \frac{\Lambda}{T} \tag{5.31}$$

where Λ is a fundamental constant (or combination of fundamental constants) with dimensions of thermodynamic temperature times wavelength. Consequently, once the constant

Λ is known, (5.31) tells us the wavelength at peak intensity λ_{max} for any system of blackbody radiation as a function of its thermodynamic temperature T.

Unfortunately, equation (5.31) is also known as *Wien's displacement law*. After all, it shows how the peak of the spectral energy density curve displaces toward shorter wavelengths (and higher frequencies) as the temperature increases. The law $\lambda_{max} = \Lambda/T$ is somewhat older than Wien's derivation of $u_\lambda(T,\lambda) = T^5 h(T\lambda)$, having been empirically discovered before 1893, even if its universal character was not immediately recognized.[13]

The displacement law $\lambda_{max} = \Lambda/T$ is particularly important because the wavelength λ_{max} at peak intensity is what we most easily observe and what we can most easily measure. And, once λ_{max} is known, we may use (5.31) to infer the temperature of an astrophysical object. This is how we know that the temperature $T[= \Lambda/\lambda_{max}]$ of the Sun's surface is approximately $5800K$.

5.8 Wien's 1896 Distribution

In 1896 Wien also proposed a particular function of the spectral energy density,

$$u_\lambda(T,\lambda) = b\lambda^{-5} e^{-a/\lambda T}, \tag{5.32}$$

known as *Wien's distribution*, that realized the form required by his displacement law (5.28). The constants a and b depend only upon fundamental dimensional constants and numerical factors and not at all upon thermodynamic variables. Wien's distribution can also be written in frequency form

$$u_\nu(T,\nu) = \frac{b\nu^3}{c^4} e^{-a\nu/cT} \tag{5.33}$$

where c is the speed of light.

Wien's distribution was, at first, found to be consistent with the available observations. Even so, (5.32) and (5.33) had only a brief influence. For while Wien's distribution was not unmotivated, these motivations were not long compelling. Nevertheless, Wien's distribution played an important role in Max Planck's search for a theoretically compelling and empirically verified derivation of the spectral energy density and in Albert Einstein's analysis of high-frequency blackbody radiation. Yet while Wien's distribution was important to Planck and Einstein, it is only Wien's displacement law that has stood the test of time.

5.9 Wien's 1893 Derivation

Appendix B contains Louis Buchholtz's translation of Wien's 1893 paper, "A New Relationship between Blackbody Radiation and the Second Law of Thermodynamics." Interestingly, Wien's derivation in this paper does not quite demonstrate what has, since 1899, been called "Wien's displacement law," stopping short of a full derivation.

Wien's derivation is excessively complex, but it seems to accomplish the following in the following order.

(1) An indirect proof that blackbody radiation, when adiabatically and reversibly compressed, remains blackbody radiation.

(2) A derivation of the electromagnetic adiabatic invariant $V^{1/3}\lambda^{-1}$. Although Wien does not use the phrase *adiabatic invariant*, it is clear that $V^{1/3}\lambda^{-1}$ is an adiabatic invariant since this quantity remains constant during a reversible, adiabatic compression "that is vanishingly slow compared to the speed of light."

(3) A derivation of the thermodynamic adiabatic invariant $uV^{4/3}$.

(4) Recognition that $u\lambda^4\,[=uV^{4/3}/(V^{1/3}\lambda^{-1})^4]$ is an adiabatic invariant.

(5) A claim that, given the Stefan-Boltzmann law, uT^{-4} is invariant.

(6) Recognition that $\lambda T[=(u\lambda^4/uT^{-4})^{1/4}]$ is an adiabatic invariant.

(7) An introduction of the spectral energy density $u_\lambda(T,\lambda)$. Then, in Wien's words, "Let us . . . imagine entering the quantity λ as abscissa and $u_\lambda(T,\lambda)$ as ordinate." Then, he imagines a reversible, adiabatic compression (or expansion) that takes a system described by $u_{\lambda_o}(T_o,\lambda_o)\,d\lambda_o$ into one described by $u_\lambda(T,\lambda)\,d\lambda$. (Note here that Wien is treating a subinterval of the spectral energy as a thermodynamic system.) Figure 5.2 illustrates these quantities and this transformation. Given the adiabatic invariant λT, the intervals $d\lambda_o$ and $d\lambda$ are related to each other by

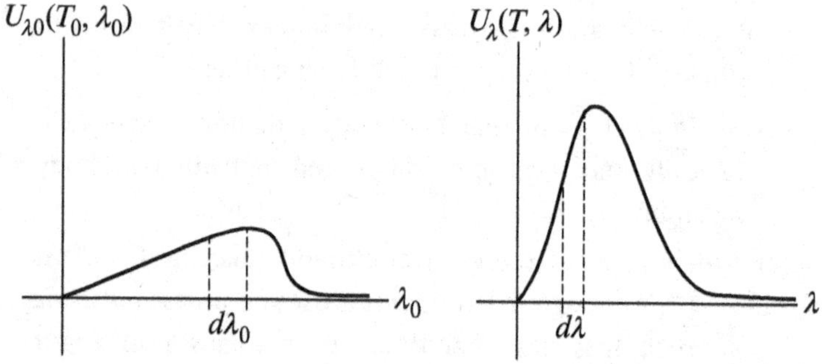

Figure 5.2.
Transformation of $u_{\lambda_o}(T_o,\lambda_o)$ into $u_\lambda(T,\lambda)$ given that $Td\lambda = T_o d\lambda_o$ and $u_\lambda(T,\lambda)/T^5 = u_{\lambda_o}(T_o,\lambda_o)/T_o^5$.

$$Td\lambda = T_0 d\lambda_0. \tag{5.34}$$

(8) The area produced by each differential interval and its respective spectral energy density must be scaled by temperature in order to produce, when integrated over all wavelengths, the Stefan-Boltzmann law. Therefore,

$$\frac{u_\lambda(T,\lambda)}{T^4} d\lambda = \frac{u_{\lambda_0}(T_0,\lambda_0)}{T_0^4} d\lambda_0 \tag{5.35}$$

or, given (5.34),

$$\frac{u_\lambda(T,\lambda)}{T^5} = \frac{u_{\lambda_0}(T_0,\lambda_0)}{T_0^5}. \tag{5.36}$$

Wien concludes his 1893 paper with (5.36). However, a few more steps are necessary in order to turn (5.36) into what has come to be known as "Wien's displacement law." In particular, note that (5.36) is equivalent to

$$u_\lambda(T,\lambda)T^{-5} = ad.inv.. \tag{5.37}$$

Furthermore, since each adiabatic invariant of a system changes if and only if every other adiabatic invariant of that system changes, we infer that a relationship exists between the two adiabatic invariants $u_\lambda(T,\lambda)T^{-5}$ and λT. Therefore,

$$u_\lambda(T,\lambda)T^{-5} = f(\lambda T), \tag{5.38}$$

where the function $f(\lambda T)$ is undetermined. Equation (5.38) is Wien's displacement law.

6 The Damped, Driven, Simple Harmonic Oscillator

6.1 Planck Resonator

According to Kirchhoff's law, the spectral energy density of blackbody radiation $u_\omega(T,\omega)$, where $\omega = 2\pi\nu$, is independent of the kind of material with which the radiation is in thermal equilibrium as long as that material is "black," that is, receptive of all frequencies of radiation incident upon it. Therefore, in investigating the form of $u_\omega(T,\omega)$ Planck had the freedom to imagine a material, composed of idealized properties that include "blackness," in thermal equilibrium with blackbody radiation. According to one of Planck's *Eight Lectures on Theoretical Physics* delivered at Columbia University in 1909,

> the nature of the bodies [resonators] . . . is therefore quite immaterial, and it is certainly immaterial whether such bodies actually exist anywhere in nature, so long as their existence and their properties are compatible throughout with the laws of electrodynamics and of thermodynamics.[1]

Consequently, Planck imagined a material composed of simple harmonic oscillators that respond to the blackbody radiation in which they are immersed. He called these oscillators *resonators* because each has a natural frequency ω_0 at which it can be made to resonate.

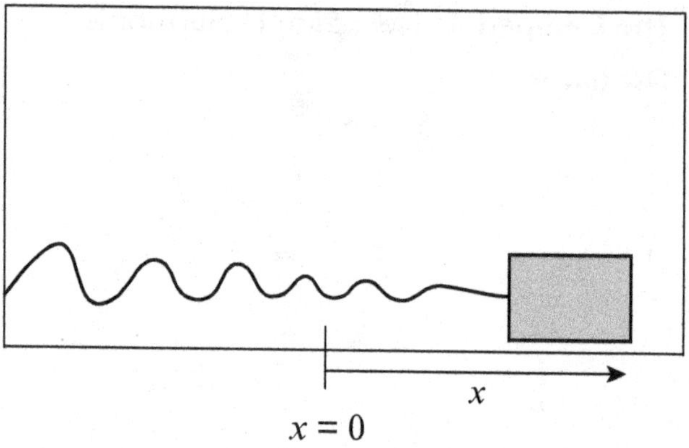

Figure 6.1
Simple harmonic oscillator.

6.2 Simple Harmonic Oscillator

We begin with a simple harmonic oscillator composed of a point mass m attached to the end of a horizontal spring that exerts a linear restoring force $-kx$ on that mass, as shown in Figure 6.1. Here x is the spring's displacement from equilibrium and k is the spring constant. Therefore, the equation of motion of this point mass is

$$m\ddot{x} = -kx \tag{6.1}$$

where $\ddot{x} = d^2x/dt^2$. Equation (6.1) is equivalent to

$$\ddot{x} + \omega_0^2 x = 0 \tag{6.2}$$

where $\omega_0 [= \sqrt{k/m}]$ is the *natural frequency* of the spring-mass system.

The simple harmonic oscillator is not merely a toy model that Planck and other physicists like to imagine and analyze. It reveals important features of all systems that oscillate around their equilibrium positions: pendulums, guitar strings, drumheads, columns of air, and any material system in stable equilibrium. This includes most of the stationary objects that surround us. For if the force on a material point is $f(x)$ where x is its displacement from its equilibrium position at $x = 0$, then the expansion of that force around $x = 0$, through two terms, is $f(x) \approx f(0) + f'(0)x$. Since the system is in equilibrium, $f(0) = 0$. And if that equilibrium is stable, then $f'(0) < 0$ and so $f(x) = -kx$ where $k = |f'(0)|$.

The linearity of this second-order, ordinary differential equation (6.2) means that its general solution is a linear combination of two independent solutions. A simple inspection reveals that two such independent solutions are $\sin(\omega_0 t)$ and $\cos(\omega_0 t)$. Therefore, the general solution of (6.2) has the form

$$x(t) = A\cos(\omega_0 t) + B\sin(\omega_0 t) \tag{6.3}$$

or, alternatively and equivalently,

$$x(t) = C\cos(\omega_0 t + \theta) \tag{6.4}$$

where the constants A and B, or, alternatively, C and θ, can be specified in terms of the initial conditions of the mass. For instance, when $x(0)$ and $\dot{x}(0)$ are the initial position and velocity of the mass, $A = x(0)$, $B = \dot{x}(0)/\omega_0$, $\theta = -\tan^{-1}[\dot{x}(0)/\omega_0 x(0)]$, and $C = \sqrt{x(0)^2 + \dot{x}(0)^2/\omega_0^2}$. Given solution (6.3) or, equivalently, (6.4), it is clear that

$$\frac{d}{dt}\left(\frac{m\dot{x}^2}{2} + \frac{kx^2}{2}\right) = 0. \tag{6.5}$$

Therefore, $m\dot{x}^2/2 + kx^2/2$ is a constant of the oscillator motion. As is traditional since Rudolf Clausius's masterly fashioning of classical thermodynamics, we use the symbol U to stand for the internal energy of the oscillator, in which case

$$U = \frac{m\dot{x}^2}{2} + \frac{kx^2}{2}. \tag{6.6}$$

6.3 The Damped, Simple Harmonic Oscillator

If the simple harmonic oscillator is submerged in a viscous fluid, a new force $-\gamma m\dot{x}$, proportional to the oscillator velocity $\dot{x}$, the oscillator mass m, and a *damping decrement* γ, retards the oscillator motion. In this case the equation of motion (6.2) becomes

$$\ddot{x}+\omega_0^2 x+\gamma\dot{x}=0. \tag{6.7}$$

Planck's resonators were not retarded with a force of this form, but because equation (6.7) is, like (6.2), a linear, second-order, ordinary differential equation, finding its solution is a gentle introduction to Planck's actual method.

A mere inspection of (6.7) does not readily suggest its solution. Instead, we look for two independent solutions of the form e^{pt} where the constant p is some function of the system parameters ω_0 and γ. In particular, solutions of the form $x = const.\cdot e^{pt}$ turn (6.7) into the *auxiliary equation*

$$p^2+\omega_0^2+\gamma p=0 \tag{6.8}$$

whose solutions (those of a quadratic) are given by

$$\begin{aligned} p_\pm &= \frac{-\gamma\pm\sqrt{\gamma^2-4\omega_0^2}}{2} \\ &= \frac{-\gamma}{2}\pm i\omega_0\sqrt{1-\frac{\gamma^2}{4\omega_0^2}}. \end{aligned} \tag{6.9}$$

Therefore, the general solution of (6.7) is

$$\begin{aligned} x(t) &= e^{-\gamma t/2}\left[A\cos\omega_0' t+B\sin\omega_0' t\right] \\ &= Ce^{-\gamma t/2}\sin(\omega_0' t+\theta) \end{aligned} \tag{6.10}$$

where

$$\omega_0' = \omega_0\sqrt{1-(\gamma/2\omega_0)^2}. \tag{6.11}$$

Here A and B, or alternatively C and θ, are constants that can be expressed in terms of the initial conditions. According to (6.10) and (6.11) this system oscillates at a reduced, natural frequency $\omega_0' \, [\le \omega_0]$ while its amplitude decreases exponentially at rate $\gamma/2$. In the long run for which $\gamma t/2 >> 1$, the oscillation amplitude decays to insignificance.

6.4 The Damped, Driven, Simple Harmonic Oscillator

We now apply a harmonic driving force $f_0 \cos(\omega t)$ to the damped, simple harmonic oscillator where the amplitude of the driving force f_0 is a constant. The complete equation of motion is now

$$m\ddot{x} = -kx - m\gamma\dot{x} + f_0 \cos(\omega t). \tag{6.12}$$

With this addition the form of the differential equation changes from one that is *homogeneous* in the dependent variable x (such that multiplying every occurrence of x by a constant leaves the equation unchanged) to one that is *inhomogeneous* in x.

Solving linear, inhomogeneous equations of the form (6.12) requires a special method—one to which we are led by the observation that the driving force keeps the amplitude of the solution $x(t)$ from decaying to zero. Therefore, in the long run, the solution of (6.12) should oscillate at the driving frequency ω rather than at the natural frequency ω_0 of the system. Therefore, we look for a particular solution of (6.12) of the form

$$x(t) = A\cos(\omega t + \phi) \tag{6.13}$$

where the oscillation amplitude A and phase ϕ are chosen so as to satisfy the equation of motion (6.12).

To characterize solution (6.13), we transform (6.12) into its equivalent

$$\ddot{x}+\omega_o^2 x+\gamma\dot{x}=\left(\frac{f_o}{m}\right)\cos(\omega t). \tag{6.14}$$

Then we calculate the time derivatives of the particular solution $x(t)$ of (6.13) and find that $\dot{x}(t)=-\omega A\sin(\omega t+\phi)$ and $\ddot{x}(t)=-\omega^2 A\cos(\omega t+\phi)$. Substituting these into (6.14) produces the requirement

$$\left(\omega_o^2-\omega^2\right)A\cos(\omega t+\phi)-\gamma\omega A\sin(\omega t+\phi)=\left(\frac{f_o}{m}\right)\cos(\omega t) \tag{6.15}$$

on the amplitude A and the phase φ. Replacing $\cos(\omega t+\phi)$ in (6.15) with its expansion $\cos(\omega t)\cos(\phi)-\sin(\omega t)\sin\phi$ and $\sin(\omega t+\phi)$ with its expansion $\sin(\omega t)\cos(\phi)+\cos(\omega t)\sin(\phi)$ and collecting the independent terms, $\cos(\omega t)$ and $\sin(\omega t)$, yields

$$\begin{aligned}&\cos(\omega t)\left[\left(\omega_o^2-\omega^2\right)A\cos(\phi)-\gamma\omega A\sin(\phi)-\frac{f_o}{m}\right]+\\&\quad\sin(\omega t)\left[-\left(\omega_o^2-\omega^2\right)A\sin(\phi)-\gamma\omega A\cos(\phi)\right]=0.\end{aligned} \tag{6.16}$$

Therefore, choosing parameters A and ϕ that characterize the solution (6.13) is equivalent to choosing these parameters so that the two bracketed terms in (6.16) vanish. For this reason, the amplitude A and the phase ϕ satisfy

$$A=\frac{f_o/m}{\left(\omega_o^2-\omega^2\right)\cos\phi-\gamma\omega\sin\phi} \tag{6.17}$$

and

$$\tan\phi=\frac{-\gamma\omega}{\left(\omega_o^2-\omega^2\right)}. \tag{6.18}$$

Using (6.18) to eliminate the phase ϕ in (6.17) produces an equation for the amplitude

$$A(\omega)=\frac{f_o/m}{\sqrt{\left(\omega_o^2-\omega^2\right)^2+\gamma^2\omega^2}} \tag{6.19}$$

and inverting (6.18) produces an equation for the phase

$$\phi(\omega) = \tan^{-1}\left[\frac{-\gamma\omega}{\left(\omega_0^2 - \omega^2\right)}\right] \tag{6.20}$$

in terms of the driving frequency ω and amplitude f_0 and the system parameters γ, m, and ω_0. This solution

$$x_p(t) = A(\omega)\cos(\omega t + \phi) \tag{6.21}$$

of the equation of motion, where the amplitude $A(\omega)$ and phase $\phi(\omega)$ are given by (6.19) and (6.20), is called a *particular solution* of the equation of motion (6.14).

The particular solution $x_p(t)$ is not the general solution of the equation of motion (6.14), since it is not a linear combination of two independent solutions joined with two constants that can be expressed in terms of two initial conditions. However, a general solution can be constructed by adding the decaying, general solution (6.10) of the homogeneous equation of motion (6.7) to the particular solution (6.21) so that

$$x(t) = Ce^{-\gamma t/2}\sin(\omega_0' t + \theta) + A(\omega)\cos(\omega t + \phi) \tag{6.22}$$

where ω_0', $A(\omega)$, and ϕ are defined by (6.11), (6.19), and (6.20). This leaves the constants C and θ to be expressed in terms of the initial conditions.

6.5 Lorentzian Approximation for Weak Damping

Planck was free to endow his resonators with weak damping, as defined by the ordering $\gamma << \omega_0$. In this regime the response function $A(\omega)$, given by (6.19), is highly peaked around the resonator's natural frequency ω_0. Furthermore, in this regime, the response curve is closely approximated, as we shall see, by expanding its denominator through second order in the small quantities $\omega/\omega_0 - 1$ and γ/ω_0. The result is

$$A(\omega)=\frac{f_o/m\omega_o^2}{\sqrt{4(1-\omega/\omega_o)^2+(\gamma/\omega_o)^2}}. \tag{6.23}$$

The square $A^2(\omega)$ of (6.23) is proportional to the *Cauchy-Lorentz* function, or more briefly a *Lorentzian*, sometimes used to model spectral line shapes. This squared function normalized in the following way

$$\frac{2\gamma\omega_o^3m^2A^2(\omega)}{f_o^2\pi}=\frac{\gamma/(2\omega_o\pi)}{(1-\omega/\omega_o)^2+(\gamma/2\omega_o)^2} \tag{6.24}$$

is symmetrically peaked around the resonance at $\omega=\omega_o$ and defined so that the area under its curve, in ω/ω_o space, is one.

In order to compare the unapproximated response function (6.19) to its Lorentzian approximation (6.23) we normalize the original response function (6.19) in the same way as that of (6.24) so that (6.19) becomes

$$\frac{2\gamma\omega_o^3m^2A^2(\omega)}{\pi f_o^2}=\frac{2\gamma/\pi\omega_o}{\left(1-\omega^2/\omega_o^2\right)^2+(\gamma/\omega_o)^2(\omega/\omega_o)^2}. \tag{6.25}$$

Figure (6.2) shows the functional form $2\gamma\omega_o^3m^2A^2(\omega)/\pi f_o^2$ defined by (6.25) (solid curve) and its Lorentzian approximation defined by (6.24) (dashed curve) for two different values of the damping. The more highly peaked functions are those for which $\gamma/\omega_o=0.04$ while those with a broader maximum are those for which $\gamma/\omega_o=0.4$. Evidently, the weaker the damping (that is, the smaller γ/ω_o), the closer together are the response function and its Lorentzian approximation and the more both curves are highly peaked and symmetric.

Because Planck's resonators are weakly damped, driven, simple harmonic oscillators, he was able to exploit the convenient features of the Lorentzian function (6.24), including its symmetry around a resonant frequency, its unit normalization, and its increasing "peakedness" as $\gamma/2\omega_o$ diminishes.

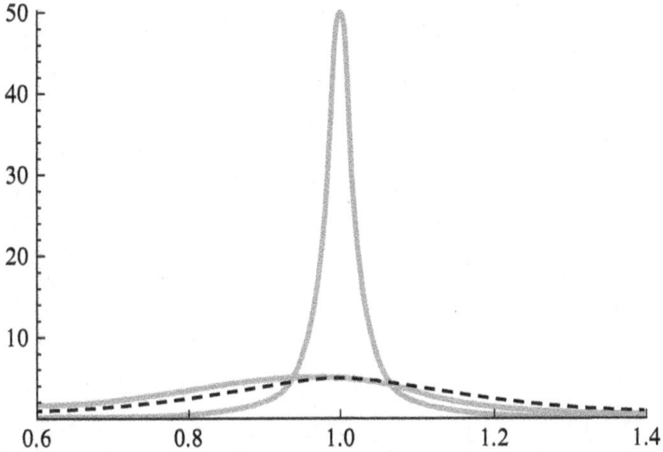

Figure 6.2
$2\gamma\omega_o^3 m^2 A^2(\omega)/\pi f_o^2$ versus ω/ω_o for the unapproximated response curve (6.25) (solid) and its Lorentzian approximation (6.24) (dashed) when $\gamma/\omega_o = 0.4$ (broadly peaked curves) and when $\gamma/\omega_o = 0.04$ (highly peaked curves). When the damping is small, the unapproximated response function and its Lorentzian approximation are indistinguishable. The curves are normalized so that the area under the Lorentzian is one.

These same features were later used by Paul Dirac to introduce his *delta function.*[2] Indeed, the limit

$$\lim_{\gamma/2\omega_o \to 0}\left[\frac{\gamma/(2\omega_o\pi)}{(1-\omega/\omega_o)^2 + (\gamma/2\omega_o)^2}\right] = \delta\left(1-\frac{\omega}{\omega_o}\right) \tag{6.26}$$

is one representation of the Dirac delta $\delta(1-\omega/\omega_o)$. Had Planck been aware of the Dirac delta he could have used it to simplify part of his derivation. Instead, Planck had to discern its properties piece by piece and resort to a workaround.

7 The Fundamental Relation

7.1 The Fundamental Relation

Planck's groundbreaking papers of 1900 and 1901 depend upon a relation that had yielded to Planck after "a long series of investigations."[1] He considered this relation,

$$u_\omega(T,\omega) = \frac{\omega^2}{\pi^2 c^3} U_1(T,\omega) \tag{7.1}$$

where $\omega = 2\pi\nu$, so important that he called it the "fundamental relation."[2] Planck's fundamental relation connects the frequency version of the spectral energy density of blackbody radiation $u_\omega(T,\omega)$ to the average energy $U_1(T,\omega)$ of a single, charged, *Planck resonator* with natural frequency ω when the two are in equilibrium with each other.

A Planck resonator is, essentially, a driven, charged, simple harmonic oscillator with a natural frequency and small radiation damping term. As such the fundamental relation follows inevitably from a classical analysis of the interaction of a resonator and its driving field. Readers of Planck's 1900 and 1901 papers in which he first derived the spectral energy density of blackbody radiation are asked to accept the fundamental relation as a given. But even if result (7.1) is an uncontroversial deduction from classical physics, its derivation is neither simple nor straightforward.

What Planck's contemporaries took for granted and what they found problematic in Planck's derivation are the reverse of what modern readers take for granted and find problematic. After all, the unusual part of Planck's analysis of blackbody radiation is his determination of how the average thermodynamic energy of a single Planck resonator $U_1(T,\omega)$ is related to the temperature T of blackbody radiation. For it was in deriving this function $U_1(T,\omega)$ that Planck was forced to introduce the universal constant h that he called the *quantum of action*. Yet today one finds that the derivation of $U_1(T,\omega)$ unfolds straightforwardly while the derivation of the fundamental relation (7.1) follows a tortuous path that combines several conceptually distinct steps and, seemingly, a number of approximations.

Our aim is to make Planck's derivation of the fundamental relation as transparent and convincing as possible and to show why this result is demanded by the principles of classical physics. To do so we lean heavily on the derivation found in the fifth of eight lectures Planck delivered at Columbia University in 1909.[3]

7.2 The Planck Resonator Model

According to Kirchhoff's law, the spectral energy density of blackbody radiation $u_\omega(T,\omega)$ is independent of the kind of material with which it is in thermal equilibrium as long as this material is black, that is, receptive of all frequencies that fall upon it. Therefore, in investigating the form of $u_\omega(T,\omega)$ Planck had the freedom to imagine a material constructed out of simple, idealized properties that included blackness. According to

Planck this material need not be realized in nature. It needs only to be consistent with the laws of nature.

Consequently, Planck chose a material composed of charged, simple harmonic oscillators that are small compared to the wavelengths of the blackbody radiation with which they interact. He called these oscillators "resonators" because they resonate with the frequencies of the blackbody radiation field close to their natural frequencies. To illustrate, we choose a single, simple harmonic oscillator composed of a point mass m and charge q fixed on the end of a spring that is aligned in the x direction and exerts a linear restoring force $-kx$ when displaced a distance x from equilibrium—as illustrated in Figure 6.1. An electric field driving force qE_x exerts a force on the oscillator. But it is important to remember that a blackbody must be composed of many such harmonic oscillators whose natural frequencies cover the spectrum of blackbody frequencies.

The resonator equation of motion is

$$m\ddot{x} + kx = qE_x \tag{7.2}$$

or, equivalently,

$$\ddot{x} + \omega_0^2 x = \frac{q}{m} E_x \tag{7.3}$$

where $\omega_0 \left[= \sqrt{k/m}\right]$ is the resonator's natural frequency. Therefore, the rate at which the electric field E_x contributes to the resonator energy $m\dot{x}^2/2 + kx^2/2$ is given by

$$\frac{d}{dt}\left(\frac{m\dot{x}^2}{2} + \frac{kx^2}{2}\right) = qE_x\dot{x}. \tag{7.4}$$

Equation (7.4) does not yet account for the energy radiated by the resonator charge as it moves back and forth.

The oscillator radiates energy because its charge accelerates along with the oscillator mass. And the *average* power P with which a charge q having acceleration a radiates energy is given by the *Larmor formula*

$$P = \frac{q^2 a^2}{6\pi\varepsilon_0 c^3} \tag{7.5}$$

where the average is over an interval that includes a number of periods τ of the electromagnetic wave that is radiated. Integrating equation (7.4) over this interval τ starting at an arbitrary time t and including a term that incorporates the radiated power P, that is, incorporates (7.5), results in

$$\frac{1}{\tau}\int_t^{t+\tau} \frac{d}{dt}\left(\frac{m\dot{x}^2}{2} + \frac{kx^2}{2} - qE_x\dot{x}\right)dt = \frac{-1}{\tau}\int_t^{t+\tau} m\sigma\ddot{x}^2 dt \tag{7.6}$$

where the quantity

$$\sigma = \frac{q^2}{6\pi\varepsilon_0 mc^3} \tag{7.7}$$

with the dimension of time simplifies our notation. The term on the right-hand side of (7.6) is further simplified by integration by parts. Thus,

$$\begin{aligned}\int_t^{t+\tau} \ddot{x}^2 dt &= \int_t^{t+\tau} \ddot{x}d\dot{x} \\ &= \ddot{x}\dot{x}\Big|_t^{t+\tau} - \int_t^{t+\tau} \dot{x}d\ddot{x} \\ &= -\int_t^{t+\tau} \dot{x}\cdot\dddot{x}dt\end{aligned} \tag{7.8}$$

where the term $\ddot{x}\dot{x}\big|_t^{t+\tau}$ vanishes because of the assumed periodicity of the oscillator. Relation (7.8) transforms (7.6) into

$$\begin{aligned}0 &= \frac{1}{\tau}\int_t^{t+\tau}\left[\frac{d}{dt}\left(\frac{m\dot{x}^2}{2} + \frac{kx^2}{2}\right) - qE_x\dot{x} - m\sigma\dot{x}\cdot\dddot{x}\right]dt \\ &= \frac{1}{\tau}\int_t^{t+\tau} \dot{x}(m\ddot{x} + kx - qE_x - m\sigma\dddot{x})dt\end{aligned} \tag{7.9}$$

from which comes the linear equation of motion

$$m\ddot{x} = -kx + qE_x + m\sigma\dddot{x} \tag{7.10}$$

or, equivalently,

$$\ddot{x} + \omega_0^2 x - \sigma\dddot{x} = \frac{q}{m}E_x \tag{7.11}$$

where $\omega_0 = \sqrt{k/m}$ is the natural frequency of the oscillator.[4]

At this point we assume that the driving electric field is a component of only one frequency, so that $E_x = E_{xo}\cos\omega t$ where ω is the driving frequency. Then the equation of motion (7.11) becomes

$$\ddot{x} + \omega_0^2 x - \sigma\dddot{x} = \frac{qE_{xo}}{m}\cos\omega t. \tag{7.12}$$

This equation, of course, concentrates on only one part of the spectrum of blackbody radiation waves that interact with each and every Planck resonator. However, solving (7.12) is only a stage on the way. For, given the linearity of (7.12), its solutions summed over the whole spectrum of driving frequencies ω compose a general solution.

Note that, except in its damping term, the equation of motion (7.12) is identical to that of the damped, driven, simple harmonic oscillator equation of motion (6.14). The two equations differ in that the radiation damping term in (7.12) is $\sigma\dddot{x}$ while the dissipation term in (6.14) is $\gamma\dot{x}$. Therefore, up to a point, our method of solving (7.12) will mimic that of solving (6.14).

7.3 The Weakly Damped Planck Resonator

In the absence of a driving force the equation of motion (7.12) becomes

$$\ddot{x} + \omega_o^2 x - \sigma \dddot{x} = 0. \tag{7.13}$$

As in section 6.3 we look for solutions of the form e^{pt}. In this way, we generate the auxiliary equation

$$p^2 + \omega_o^2 - \sigma p^3 = 0 \tag{7.14}$$

for the index p. Planck resonators are endowed with weak damping. In this context "weak" means that

$$\sigma p^3 << p^2, \omega_o^2 \tag{7.15}$$

and, therefore, to zero order in an expansion in the small term σp

$$p = \pm i\omega_o. \tag{7.16}$$

And through first order in the small term σp the auxiliary equation (7.14) becomes

$$p^2 + \omega_o^2 \pm i\sigma\omega_o^3 = 0. \tag{7.17}$$

The solution of (7.17) is

$$\begin{aligned} p &= \pm i\sqrt{\omega_o^2 \pm i\sigma\omega_o^3} \\ &= \pm i\omega_o\sqrt{1 \pm i\sigma\omega_o} \\ &= \pm i\omega_o\left(1 \pm \frac{i\sigma\omega_o}{2}\right) \\ &= \pm i\omega_o - \frac{\sigma\omega_o^2}{2} \end{aligned} \tag{7.18}$$

since $\sigma\omega_o << 1$ is required by the weak damping condition (7.15). (Note that $(\pm 1)^3 = \pm 1$ and $i^3 = -i$.) Therefore, the general solutions of the weakly damped Planck resonator decay exponentially in time t as $e^{-\sigma\omega_o^2 t/2}$. This kind of weak damping, in which $\sigma p << 1$ or equivalently $\sigma\omega_o << 1$, insures that the time required for one exponential decay, $2/\sigma\omega_o^2$, is long compared to the oscillation period ω_o^{-1}.

7.4 The Damped, Driven Planck Resonator

As in section 6.4 we solve the damped, driven equation of motion (7.12) for a particular solution

$$x_p(t) = A\cos(\omega t + \phi) \tag{7.19}$$

that will persist long after the initial conditions have damped away. The amplitude A and phase ϕ that define this particular solution are found in the same way that (6.19) and (6.20) were found in section 6.4. Accordingly,

$$A(\omega) = \frac{qE_{xo}/m}{\sqrt{\left(\omega_o^2 - \omega^2\right)^2 + \left(\sigma\omega^3\right)^2}} \tag{7.20}$$

and

$$\phi(\omega) = \tan^{-1}\left[\frac{-\sigma\omega^3}{\omega_o^2 - \omega^2}\right]. \tag{7.21}$$

In the weak damping regime, $\sigma\omega_o << 1$, this amplitude $A(\omega)$ is highly peaked around the natural frequency ω_0 of the resonator. To see this we expand the denominator of (7.20) through second order in the small quantities $\omega/\omega_o - 1$ and $\sigma\omega_o$. Then the response function $A(\omega)$ becomes

$$A(\omega) = \frac{qE_{xo}/m}{\sqrt{4\omega_o^2\left(\omega_o - \omega\right)^2 + \left(\sigma\omega_o^3\right)^2}}, \tag{7.22}$$

that is, it becomes proportional to a Lorentzian, which in the limit of vanishing damping becomes proportional to a Dirac delta.

The average energy of a single oscillator with natural frequency ω_0 driven by an electric field with frequency ω, given (7.19) and (7.22), is represented by

$$\begin{aligned}\overline{\frac{m\dot{x}^2}{2}+\frac{kx^2}{2}} &= \frac{mA^2\omega^2}{2}\overline{\sin^2(\omega t+\phi)}+\frac{kA^2}{2}\overline{\cos^2(\omega t+\phi)} \\ &= \frac{m\left(\omega^2+\omega_o^2\right)A^2}{4} \\ &= \frac{m\left(\omega^2+\omega_o^2\right)}{4}\frac{q^2E_{xo}^2/m^2}{\left[4\omega_o^2(\omega_o-\omega)^2+\left(\sigma\omega_o^3\right)^2\right]}\end{aligned} \tag{7.23}$$

where the bar indicates an average over a period ω^{-1} of the driving frequency and, in the last step, we have used the Lorentzian approximation (7.22) to the response function $A(\omega)$. As such (7.23) is merely the material out of which we will, in the next section, compose the energy of a Planck resonator in response to the entire spectrum of blackbody frequencies.

7.5 Resonator Responding to a Spectrum

Each side of equation (7.23) for the average energy of a single Planck resonator in terms of the frequencies ω and ω_o and amplitude E_{xo} of the driving electric field must be modified in order to transform it into a relation between the average energy of a system of identical Planck resonators driven by a complete spectrum of blackbody radiation. The signifier of thermal equilibrium, a common temperature T, will characterize both the average resonator energy and the blackbody radiation.

We first recognize that, while the Planck resonator described by (7.23) is driven by a single component $E_{xo}\cos\omega t$ of the electric field, this component is only one of six components of an electromagnetic wave each of which contributes equally to the local energy density. Therefore, this energy density $u_\omega(T,\omega)$ per driving frequency between ω and $\omega+d\omega$ is related to the electric field amplitude of the driving field by the following equivalences

$$u_\omega(T,\omega)d\omega \text{ is equivalent to } 6\cdot\left(\frac{\varepsilon_o E_{xo}^2}{2}\right)\overline{\cos^2 \omega t}$$
$$\text{is equivalent to } \frac{3}{2}\varepsilon_o E_{xo}^2. \tag{7.24}$$

Therefore,

$$\frac{2}{3\varepsilon_o}u_\omega(T,\omega)d\omega \text{ is equivalent to } E_{xo}^2 \tag{7.25}$$

will be used in the right-hand side of (7.23).

Secondly, we represent the average energy of the resonators with a natural frequency of ω_o driven by blackbody radiation frequencies within an interval between ω and $\omega + d\omega$ by $U_{1,\omega}(T,\omega)d\omega$ where $U_{1,\omega}(T,\omega)$ is the *energy density per unit frequency interval* of these resonators. Therefore,

$$U_{1,\omega}(T,\omega)d\omega \text{ is equivalent to } \overline{\frac{m\dot{x}^2}{2}+\frac{kx^2}{2}} \tag{7.26}$$

will be used on the left-hand side of (7.23).

Together equivalences (7.25) and (7.26) transform (7.23) into the differential equation

$$U_{1,\omega}(T,\omega)d\omega = \frac{q^2\left(\omega^2+\omega_o^2\right)}{4m}\left(\frac{2}{3\varepsilon_o}\right)\frac{u_\omega(T,\omega)d\omega}{\left[4\omega_o^2\left(\omega_o-\omega\right)^2+\left(\sigma\omega_o^3\right)^2\right]}, \tag{7.27}$$

from whence comes our main result. Integrating both sides of (7.27) over all driving frequencies from $\omega = 0$ to $\omega = \infty$ yields, on the left-hand side, the average energy of a resonator with natural frequency ω_o because

$$U_1(T,\omega_o) = \int_0^\infty U_{1,\omega}(T,\omega)d\omega. \tag{7.28}$$

Therefore, this integration of (7.27) produces

$$U_1(T,\omega_o) = \frac{q^2}{6m\varepsilon_o} \int_0^\infty \frac{(\omega^2+\omega_o^2)u_\omega(\omega,T)d\omega}{\left[4\omega_o^2(\omega_o-\omega)^2+(\sigma\omega_o^3)^2\right]}$$
$$= \frac{q^2\pi}{12m\varepsilon_o\omega_o^4\sigma} \int_0^\infty (\omega^2+\omega_o^2)u_\omega(\omega,T)\delta_L\left(1-\frac{\omega}{\omega_o}\right)d\left(\frac{\omega}{\omega_o}\right) \tag{7.29}$$

where the symbol $\delta_L(1-\omega/\omega_o)$ represents the normalized Lorentzian defined by

$$\delta_L\left(1-\frac{\omega}{\omega_o}\right) = \frac{\sigma\omega_o/2\pi}{(1-\omega/\omega_o)^2+(\sigma\omega_o/2)^2}. \tag{7.30}$$

In the limit of vanishingly small $\sigma\omega_o/2$, the normalized Lorentzian $\delta_L(1-\omega/\omega_o)$ becomes the Dirac delta $\delta(1-\omega/\omega_o)$. Consequently, this limit reduces (7.29) to

$$U_1(T,\omega_o) = \frac{q^2\pi}{6m\varepsilon_o\omega_o^2\sigma}u_{\omega_o}(T,\omega_o). \tag{7.31}$$

Given the definition (7.7) of the coefficient of damping σ, equation (7.31) is equivalent to

$$U_1(T,\omega_o) = \frac{\pi^2c^3}{\omega_o^2}u_{\omega_o}(T,\omega_o). \tag{7.32}$$

On rearranging terms and replacing ω_o with ω, (7.32) is expressed as

$$u_\omega(T,\omega) = \frac{\omega^2}{\pi^2c^3}U_1(T,\omega), \tag{7.33}$$

which is the fundamental relation.

Planck was pleased because this result (7.33) contained no parameters peculiar to the resonator model—neither its mass m, nor its spring constant k, nor its charge q, nor its damping interval σ. The only remnant of the resonator is its ability to resonate with a frequency ω of the blackbody radiation field.

Note that because the spectral energy density $u_\omega(T,\omega)$ is a continuous function of ω, this radiation must be in equilibrium with a collection of Planck resonators each having a different natural frequency ω that as a whole are continuously distributed. This is, of course, only to say that the collection of resonators must compose a blackbody.

We will find it convenient to cast the fundamental relation (7.33) in terms of the frequency $\nu[=\omega/2\pi]$ via $u_\nu d\nu = u_\omega d\omega$. Thus,

$$u_\nu(T,\nu) = \frac{8\pi\nu^2}{c^3} U_1(T,\nu). \tag{7.34}$$

In deriving the fundamental relation Planck traded one problem for a supposedly easier one. In place of finding the spectral energy density function $u_\nu(T,\nu)$ Planck's task was now to find the average energy $U_1(T,\nu)$ of a single resonator with natural frequency ν in thermal equilibrium with blackbody radiation at temperature T.

8 Planck's Zeroth Derivation, 1900

8.1 The Zeroth Derivation

Berlin, in the late nineteenth and early twentieth centuries, was the center of the physics world. Albert Einstein, Max von Laue, Max Planck, and Willy Wien, just to name those who would be Nobel Prize winners, worked in or near Berlin. Many of Planck's colleagues were aware of the phenomenon and the challenge of blackbody radiation. Possibly for this reason Planck published his ideas hurriedly, in increments, as they came to him. Consequently, making sense of the derivations of the spectral energy density of blackbody radiation Planck put together in 1900–1901 is a task that requires studying chains of references that link his various talks and papers into a coherent whole. The task is yet more difficult because some of Planck's ideas did not become fully transparent until years later when he had the leisure to explain his reasoning more fully.[1]

Planck's thinking reached a noteworthy, if short-lived, conclusion on October 19, 1900.[2] At that time he reported to the German Physical Society, at its fortnightly meeting, on his effort "to construct completely arbitrary expressions for the entropy" that "seemed to satisfy all requirements of the thermodynamic and electromagnetic theory" of blackbody radiation. In a later reflection on his earlier work Planck recalled that "this radiation law was an interpolation formula found by happy guesswork."[3]

Planck had learned within the last few days before his presentation that Wien's distribution failed to describe long-wavelength, blackbody radiation. And yet he had earlier and incorrectly claimed that "Wien's law [his distribution] must be necessarily true."[4] Planck's purpose on October 19 was both to explain this earlier, mistaken claim and to make amends for it by offering a better law. His remarks concluded with the appeal that "I should therefore be permitted to draw your attention to this new formula."[5]

Planck's derivation of a new formula for the spectral energy density of blackbody radiation was guided both by his attraction to theoretical simplicity and his desire for empirical confirmation. Planck's new distribution was consistent with all the available blackbody data—both the older short-wavelength data and the newer long-wavelength data—and has come down to us intact as Planck's spectral energy density of blackbody radiation. Nevertheless, his derivation of October 19, 1900 was built upon arbitrary foundations. And, while he was satisfied with its result, he quickly abandoned its derivation. Still, it was important to Planck and it is important to us—for the derivation illustrates his preoccupation with the entropy function and with irreversible processes. Some even celebrate this derivation as the beginning of the quantum revolution.[6]

Historians aggregate Planck's later derivations of the spectral energy density (of December 14, 1900 and January 1901) into Planck's *first derivation*—the first derivation that introduces quanta. This numbering makes the derivation of October 19, 1900 Planck's *zeroth derivation.*

We will present Planck's zeroth derivation in section 8.5 and, in the process, include steps that Planck took for granted. But first we review, in section 8.2, the thermodynamic requirements of Planck's system of resonators and, in sections 8.3 and 8.4, Planck's faulty derivation of the (incomplete) Wien distribution. However, this faulty derivation, completed in March of 1900, paved the way for his "happy guesswork" on October 19, 1900 as presented in section 8.5.

8.2 The Thermodynamics of Planck Resonators

Recall that, according to Kirchhoff's theorem, the spectral energy density $u_\nu(T,\nu)$ does not depend upon the material with which the blackbody radiation is in thermal equilibrium. As long as that material receives all the radiation that falls upon it, the material is a blackbody. Kirchhoff's theorem gave Planck the freedom to choose a convenient model material to be in equilibrium with the radiation as long as the model's behavior was consistent with the laws of physics. Planck's choice was a material composed of *resonators* or, as we would now say, *simple harmonic oscillators.*

To Planck these resonators constituted a thermodynamic system characterized by energy, entropy, and temperature as well as frequency. Planck had already discovered the link, based upon classical principles (the fundamental relation), between the average energy $U_1(T,\nu)$ of a single resonator and the spectral energy density $u_\nu(T,\nu)$ of the radiation with which it is in thermodynamic equilibrium, namely

$$u_\nu(T,\nu) = \frac{8\pi\nu^2}{c^3} U_1(T,\nu) \tag{8.1}$$

where ν is the natural frequency of the resonator and T is the common temperature of a collection of these resonators and the radiation field. Therefore, Planck's immediate goal was to find an expression for the average entropy $S_1(U_1)$ of a single resonator in terms of its average energy U_1 so that through $T^{-1} = \partial S_1(U_1)/\partial U_1$ he might express the energy U_1 in terms of its temperature T and its natural frequency ν. Using this expression and the fundamental relation (8.1) Planck could then determine the spectral energy density function $u_\nu(T,\nu)$.

To recreate Plank's thinking, imagine a system of N identically constructed resonators each with the same natural frequency ν. These resonators are fixed in place and are independent of one another. They are also immersed in blackbody radiation characterized by a temperature T. The resonators can only transfer energy to and receive energy from the radiation field with which they are in thermal equilibrium.

The total internal energy U_{tot} of these N resonators is an additive variable. Therefore,

$$U_{tot} = U_1 + U_2 + U_3 \ldots U_N \tag{8.2}$$

where U_i is the average internal energy of the ith resonator. Since the resonators are identically constructed and in thermodynamic equilibrium, it must be that $U_1 = U_2 = U_3 = \ldots U_N$. In this case, the additivity requirement (8.2) reduces to

$$U_{tot} = NU_1. \tag{8.3}$$

The entropy $S_{tot}(U_{tot})$ of N resonators is also an additive variable so that

$$S_{tot}(U_{tot}) = S_1(U_1) + S_2(U_2) + S_3(U_3) + \ldots S_N(U_N). \tag{8.4}$$

Again, because the resonators are identical, it must be that $S_1(U_1) = S_2(U_2) = S_3(U_3) = \ldots$ or, equivalently, $S_1(U_1) = S_2(U_1) = S_3(U_1) = \ldots$. Therefore,

$$S_{tot}(U_{tot}) = NS_1(U_1) \tag{8.5}$$

or, given (8.3),

$$S_{tot}(NU_1) = NS_1(U_1). \tag{8.6}$$

These results are simply expressions of the principle that energy and entropy are additive variables—a principle that we have, and Planck had, no reason to doubt—and the assumptions that these resonators are identically constructed and in thermodynamic equilibrium, as Planck was free to suppose. Indeed, these results follow from the very structure of thermodynamics and the flexibility one has in constructing a thermodynamic system out of identical parts.

It is clear that Planck's analysis extended as far as recognizing the truth of (8.5) and (8.6). However, had he also recognized that when $N = 1$ (8.6) reduces to

$$S_{tot}(U_1) = S_1(U_1), \tag{8.7}$$

then (8.6) could have been written as

$$S_{tot}(NU_1) = NS_{tot}(U_1). \tag{8.8}$$

This result (8.8) expresses the *extensivity* of Planck's system of identical resonators. As we shall soon see, if he had known of the extensivity (8.8) of his system and so had made use of (8.7) and (8.8), he could have avoided his incorrect derivation of the Wien distribution that we outline in the next section.

We now know, in H. B. Callen's words, that "Extensive parameters play a key role throughout thermodynamic theory."[7] However, in 1900 Clausius's mathematical expression of classical thermodynamics was only 35 years old and Clausius had left some aspects of the subject underdeveloped. While Planck understood classical thermodynamics better than most of his colleagues, he seemed unfamiliar with the language and the importance of the concept of extensivity. For this reason and in order to reproduce Planck's thinking, we (counterfactually) retain in section 8.3 the notational and the supposed substantive distinction between the two functions $S_{tot}(U_1)$ and $S_1(U_1)$.

8.3 An Irreversible Process and an Incorrect Deduction

Planck's method of deriving expressions for the average entropy of a single resonator $S_1(U_1)$ and the total entropy $S_{tot}(NU_1)$ of a system of identical resonators was to compare how a single resonator and how a group of resonators return to equilibrium after each is disturbed. However, in doing so he made a mistake that doomed his effort. Still, Planck's faulty derivation illustrates his notable effort to approach blackbody radiation through the entropy function and through the process of irreversible relaxation toward equilibrium.[8] Those wishing to focus on Planck's correct and enduring contribution should skip to section 8.5.

First, Planck imagined that an increment of energy is instantaneously added to a single resonator that was previously in equilibrium with the blackbody radiation field. This displaces the resonator from equilibrium. What will happen next? Planck, whose PhD thesis had been on the second law of thermodynamics and its relation to entropy increase, had no doubt. The energy of the resonator U_1 would decrease until it returned to its equilibrium value $U_{1,eq}$, while at the same time its entropy $S_1(U_1)$ would increase until it returned to its maximum (or equilibrium) value $S_1(U_{1,eq})$ along a path like that in Figure 8.1.

The shape of the single-resonator entropy function $S_1(U_1)$ near $U_1 = U_{1,eq}$ can be described by expanding $S_1(U_1)$ in terms of small departures from the equilibrium energy $U_{1,eq}$, that is,

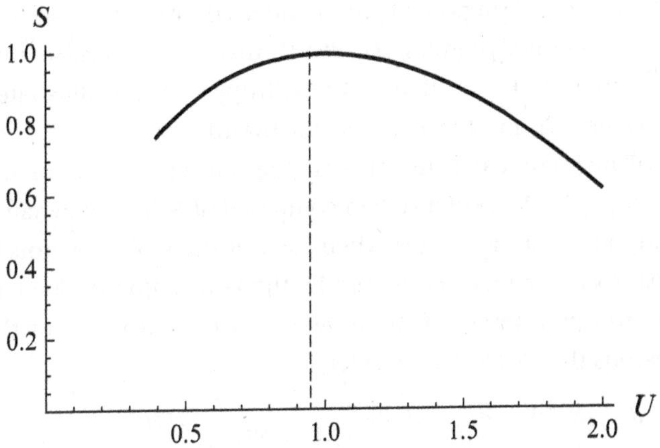

Figure 8.1

$S_1(U_1)$ versus U_1. The dashed vertical line indicates equilibrium values.

$$S_1(U_1) = S_1(U_{1,eq}) + \left.\frac{\partial S_1(U_1)}{\partial U_1}\right|_{eq}(U_1 - U_{1,eq}) + \left.\frac{\partial^2 S_1(U_1)}{\partial U_1^2}\right|_{eq}\frac{(U_1 - U_{1,eq})^2}{2} + \ldots. \tag{8.9}$$

Of course, $U_{1,eq}$ is an equilibrium energy, therefore $\partial S_1(U_1)/\partial U_1|_{U_{1,eq}} = 0$. In this case, (8.9) becomes

$$S_1(U_1) = S_1(U_{1,eq}) + \left.\frac{\partial^2 S_1(U_1)}{\partial U_1^2}\right|_{eq}\frac{(U_1 - U_{1,eq})^2}{2} + \ldots. \tag{8.10}$$

Taking the time derivative of both sides of (8.10) produces

$$\frac{dS_1(U_1)}{dt} = \left.\frac{\partial^2 S_1(U_1)}{\partial U_1^2}\right|_{eq}(U_1 - U_{1,eq})\frac{dU_1}{dt}, \tag{8.11}$$

which describes the relationship between the time derivatives $dS_1(U_1)/dt$ and dU_1/dt. Since, if $U_1 - U_{1,eq} > 0$ and

$dU_1/dt < 0$ (as supposed earlier) and according to the second law of thermodynamics, $dS_1/dt > 0$, therefore it must be that $[\partial^2 S_1(U_1)/\partial U_1^2]_{eq} < 0$; that is, the entropy curve, as illustrated in Figure 8.1, must be concave downward.

Planck compared (8.11) with the rate of increase of the entropy $S_{tot}(NU_1)$ of a system composed of N thermodynamically identical resonators when each of these N resonators is disturbed from its equilibrium by the same amount. The rate of entropy increase of the system of N resonators is, for the reasons that led to (8.11), given by

$$\begin{aligned}\frac{dS_{tot}(NU_1)}{dt} &= \left.\frac{\partial^2 S_{tot}(NU_1)}{\partial(NU_1)^2}\right|_{eq}\left(NU_1 - NU_{1,eq}\right)\frac{d(NU_1)}{dt}\\ &= N^2\left.\frac{\partial^2 S_{tot}(NU_1)}{\partial(NU_1)^2}\right|_{eq}\left(U_1 - U_{1,eq}\right)\frac{dU_1}{dt}.\end{aligned} \tag{8.12}$$

Planck then correctly deduced from (8.6), that is, from $S_{tot}(NU_1) = NS_1(U_1)$, that

$$\frac{dS_{tot}(NU_1)}{dt} = N\frac{dS_1(U_1)}{dt}. \tag{8.13}$$

Using (8.11) and (8.12) to substitute for both sides of (8.13) one arrives at a relation,

$$N\left.\frac{\partial^2 S_{tot}(NU_1)}{\partial(NU_1)^2}\right|_{eq} = \left.\frac{\partial^2 S_1(U_1)}{\partial U_1^2}\right|_{eq}, \tag{8.14}$$

between the second derivatives $\partial^2 S_{tot}(NU_1)/\partial(NU_1)^2$ and $\partial^2 S_1(U_1)/\partial U_1^2$ evaluated at the equilibrium energy $U_{1,eq}$.

At this point Planck made a mistake—perhaps an understandable mistake. He incorrectly assumed that the two functions, $\partial^2 S_{tot}(NU_1)/\partial(NU_1)^2|_{eq}$ and $\partial^2 S_1(U_1)/\partial U_1^2|_{eq}$, are the same function of their arguments so that (8.14) is equivalent to

$$Nf(NU_1) = f(U_1) \quad \text{(incorrect)} \tag{8.15}$$

where $f(x) = -\partial^2 S(x)/\partial x^2$. The functional form

$$f(x) = \frac{\alpha}{x} \tag{8.16}$$

solves (8.15) where α is an unknown constant independent of thermodynamic variables. This means that

$$\frac{\partial^2 S_1(U_{1,eq})}{\partial U_{1,eq}^2} = -\frac{\alpha}{U_{1,eq}}, \quad \text{(incorrect)} \tag{8.17}$$

which, as indicated, is incorrect.

Planck's mistake was to ignore the possibility that the entropy $S_{tot}(NU_1)$ of the N-resonator system could be an extensive function of its argument so that $S_{tot}(NU_1) = NS_{tot}(U_1)$ or, more particularly so that (8.6) obtains, $S_{tot}(NU_1) = NS_1(U_1)$. But $S_{tot}(NU_1) = NS_1(U_1)$ reduces (8.14) to an identity that tells us nothing new. Thus, while Planck's solution (8.16) $f(x) = \alpha/x$ satisfies his assumed requirement (8.15), that is, $Nf(NU_1) = f(U_1)$, this requirement (8.15) does not follow from (8.14).

8.4 The Wien Distribution

As it was, Planck proceeded by integrating the second derivative of the supposed entropy of a single resonator in (8.17) twice and found that

$$S_1(U_1) = -\alpha U_1 \ln U_1 + (\alpha + \beta)U_1 + \gamma \quad \text{(incorrect)} \tag{8.18}$$

where α, β, and γ are constants of the system. (Here we have replaced the equilibrium energy $U_{1,eq}$ with the simpler notation U_1.) The temperature T is related to the energy of a resonator by $T^{-1} = \partial S_1(U_1)/\partial U_1$. In this way, (8.18) produces

$$\frac{1}{T} = \beta - \alpha \ln U_1. \quad \text{(incorrect)} \tag{8.19}$$

Therefore,

$$U_1 = e^{\beta/\alpha} e^{-1/\alpha T}. \quad \text{(incorrect)} \tag{8.20}$$

Given the (classical) fundamental relation $u_\nu = (8\pi\nu^2/c^3)U_1$, the spectral energy density following from (8.20) is expressed by

$$u_\nu(T,\nu) = \frac{8\pi\nu^2}{c^3} e^{\beta/\alpha} e^{-1/\alpha T}. \quad \text{(incorrect)} \tag{8.21}$$

When α and β are chosen to conform with the Wien displacement law, $u_\nu(T,\nu) = \nu^3 \Theta(\nu/T)$, equation (8.21) becomes

$$u_\nu(T,\nu) = b\nu^3 e^{-a\nu/T}, \quad \text{(incorrect)} \tag{8.22}$$

where a and b are functions of fundamental dimensional constants. This is Wien's distribution discussed in section 5.8.

Of course, Planck could have worked backward from Wien's distribution (8.22) to determine that

$$\frac{\partial^2 S_1(U_{1,eq})}{\partial U_{1,eq}^2} = -\frac{\alpha}{U_{1,eq}} \quad \text{(incorrect)} \tag{8.23}$$

since at the time of this derivation experimental evidence favored Wien's distribution. And, by Planck's admission, he did just that.[9]

8.5 Planck's "Lucky Intuition"

By October 19, 1900 Planck had understood that Wien's distribution was empirically insufficient especially at long wavelengths, and, therefore, that there must be something wrong with his derivation. Since Planck's attention focused on the crucial importance of the second derivative of the entropy with respect to the energy of a resonator, his response was to improve on Wien's distribution by postulating the "simplest of all expressions for $\partial^2 S_1(U_1)/\partial U_1^2$ that leads to S_1 as a function of U_1—which is suggested by probability considerations—and which for small values of U_1 leads to Wien's expression."[10] In our notation, Planck's "simplest of all expression" is

$$\frac{\partial^2 S_1(U_1)}{\partial U_1^2} = \frac{-\alpha}{U_1(U_1+\beta)}. \tag{8.24}$$

Integrating (8.24) twice one arrives at

$$S_1(U_1) = -\alpha\left(\frac{U_1}{\beta}\right)\ln\left(\frac{U_1}{\beta}\right) + \alpha\left(1+\frac{U_1}{\beta}\right)\ln\left(1+\frac{U_1}{\beta}\right) + \gamma. \tag{8.25}$$

The definition of the inverse thermodynamic temperature $T^{-1}[=\partial S_1/\partial U_1]$ then produces

$$U_1 = \frac{\beta}{e^{\beta/\alpha T}-1}. \tag{8.26}$$

When α and β are chosen in conformity with Planck's fundamental relation $u_\nu(T,\nu) = (8\pi\nu^2/c^3)U_1(T,\nu)$ and Wien's displacement law $u_\nu(T,\nu) = \nu^3\Theta(T/\nu)$ where the function $\Theta(T/\nu)$ is undetermined, it follows that (8.26) produces (in terms of modern notation)

$$U_1 = \frac{h\nu}{e^{h\nu/kT}-1} \tag{8.27}$$

and yields the spectral energy density

$$u_\nu(T,\nu) = \frac{8\pi\nu^2}{c^3}\frac{h\nu}{e^{h\nu/kT}-1}. \tag{8.28}$$

Equation (8.28) is the correct frequency version of Planck's blackbody radiation law.

Planck's remark that his zeroth derivation as summarized in equations (8.24)–(8.28) is "suggested by probability considerations" is itself curious because he described no such probability considerations on October 19, 1900. Rather, this remark anticipates his later derivation, his "first derivation" of December 1900 and January 1901 as described in chapter 10.

Note that, in the limit of short wavelengths or high frequencies so that $h\nu/\alpha T >> 1$, Planck's spectral energy density reduces to

$$u_\nu(T,\nu) = \frac{8\pi h\nu^3}{c^3} e^{-h\nu/kT}, \tag{8.29}$$

which reproduces Wien's distribution (8.22). And, in the limit of long wavelengths or low frequencies so that $h\nu/\alpha T << 1$, Planck's spectral energy density (8.28) reduces to

$$u_\nu(T,\nu) = \frac{8\pi\nu^2}{c^3} kT. \tag{8.30}$$

This form (8.30) coincides with what is now known as the *Rayleigh-Jeans law*—a law that Lord Rayleigh (J. W. Strutt) (1842–1919) and James Jeans (1877–1946) derived from purely classical considerations. All three of the putative forms of the spectral energy density—Planck's law (8.28), the Wien distribution (8.29), and the Rayleigh-Jeans law (8.30)—satisfy the Wien displacement law. While Planck evidently knew of Rayleigh's (1900) justification of the limiting result (8.30), he paid it little attention. He was instead more impressed with recent experiments showing that, in the low-frequency, long-wavelength regime, the spectral energy density is directly proportional to the thermodynamic temperature T.[11]

8.6 Planck's New Task

On the morning of October 20, 1900 Planck's colleague Heinrich Rubens told Planck that his new distribution coincided with measurements at every point. Planck was gratified. Even so, he was not satisfied with his (zeroth) derivation. Consequently, "I began to devote myself to the task of investing it [the Planck spectral energy density] with true physical meaning."[12]

9 Boltzmann's Statistical Mechanics

9.1 Boltzmann's Physics

Ludwig Boltzmann's celebrated 1877 paper "On the Relationship between the Second Fundamental Theorem of the Mechanical Theory of Heat and Probability Calculations Regarding the Conditions for Thermal Equilibrium"[1] proposes several foundational ideas: (1) that each microscopic description or state of an isolated mechanical system consistent with thermodynamic constraints (i.e., a given internal energy U, a given volume V, etc.) is equally probable,[2] (2) that thermodynamic equilibrium corresponds to the macroscopic state consistent with the largest number of such microscopic states, and (3) that the logarithm of the number of possible microscopic states is proportional to the system entropy.

The "second fundamental theorem of the mechanical theory of heat" is, of course, the second law of thermodynamics. And the proportionality between the logarithm of the number of microscopic states of a system and its entropy proved crucial for Planck.

While Boltzmann's proposals were not of the kind that could be proven, they were physically motivated. To see this, consider a system in thermal equilibrium (that is, characterized by a single temperature) composed of a gas of identical molecules with different densities in two different compartments that are separated by a removable barrier. When the barrier separating the two compartments is removed, the system reaches a new equilibrium in which the density of the gas is uniform throughout. In this process two things happen: (1) the entropy of the system increases and (2) the number of states available to the molecules that compose the system increases. For after the barrier is removed each molecule can occupy positions throughout a larger volume. Thus, the second law of thermodynamics implied to Boltzmann that removing a constraint on a system allows its entropy to increase.

For this reason, Boltzmann saw a connection between the number P of microscopic states that a system can occupy and its entropy S. (Here we adopt Boltzmann's notation in which P stands for *permutability*, that is, the "number of permutations" or, equivalently, the number of distinct microscopic states of the system—however that might be determined. Today, we might use the symbol Ω in place of P and the word *multiplicity* in place of the word *permutability*.) Because S and P increase together, Boltzmann believed that the entropy S is a monotonically increasing function $S(P)$ of the permutability P.

Central to discovering the form of $S(P)$ is Boltzmann's assertion that the number P of distinct microscopic states available to a system is a countable number—an assertion that he made good by inventing descriptions of microstates that could, indeed, be counted. Boltzmann then discovered that maximizing $\log P$ (a real number) in place of P (an integer) allowed him to more conveniently determine the distribution that corresponds to the macrostate with the largest number of microstates. In this way, Boltzmann discovered that the logarithm

$\ln P$ of the number of microscopic states P available to an ideal monatomic gas had the same dependence on equilibrium variables—in particular, the dependence $\log(VT^{3/2})$ where V is the gas volume and T is its absolute temperature—as did the entropy S of an ideal monatomic gas.[3] This meant that $S \propto \log P$ could be used to generate the equations of state of an ideal gas.

Boltzmann even claimed that this procedure (of first counting P, then determining $\log P$, and then setting S proportional to $\log P$) applies to liquids and solids, even if these cases "seem to present extraordinary difficulties."[4] This claim took some courage, for Boltzmann had not demonstrated that $S \propto \log P$ applied more widely than to an ideal gas. Neither did he note the existence of a proportionality constant between S and $\log P$. Whatever its limitations, Boltzmann's analysis was the first successful *statistical*, that is, probabilistic, theory of entropy.

9.2 Boltzmann's Legacy

Unfortunately, Boltzmann's 1877 paper on entropy and probability is full of lengthy digressions and poorly identified methods. Maxwell once said of Boltzmann that "He could not understand me on account of my shortness, and his length was and is an equal stumbling block to me."[5]

By 1900 Planck had studied Boltzmann's 1877 paper and found in it a key to solving the problem of deriving the spectral energy density of blackbody radiation from fundamental principles.[6] Indeed, what Planck took from Boltzmann's analysis was the formulation $S \propto \log P$. But since Planck had already recognized the role of a new fundamental constant with the dimension of thermodynamic entropy, he in fact used that constant to transform $S \propto \log P$ into $S = k \log P$. The symbol k

that Planck chose for this fundamental constant stands for the German word *Konstante* and was first identified as "Planck's constant," but, of course, is now known as *Boltzmann's constant.*[7]

It is not surprising that Planck, who was the first, after Boltzmann, to use the latter's statistical approach to entropy,[8] set about trying to reconstruct and formalize Boltzmann's method. In a retrospective account,[9] Planck further transformed $S = k \log P$ into $S = k \log W$ where W stands for the German word *Wahrscheinlichkeit* meaning *probability*. However, Planck was unclear about just how this probability W was to be normalized, in contrast to Boltzmann for whom probabilities were frequencies, that is, numbers between 0 and 1. Accordingly, $S = k \log W$ would have meant to Boltzmann that the entropy of a system is a negative number.

To summarize, Planck's reformulation of Boltzmann's proportionality $S \propto \log P$ as $S = k \log W$ goes beyond Boltzmann's contribution in at least two ways: (1) in adopting the proportionality constant k and (2) in replacing the permutability P with the probability W. Thus, it is ironic that Planck's equation $S = k \log W$ rather than Boltzmann's proportionality $S \propto \log P$ appears on the tombstone erected at Boltzmann's grave in 1933 some 27 years after his death by suicide in 1906.

9.3 Boltzmann's First Calculation

Boltzmann's 1877 paper is organized into several model calculations and their application to ideal gases. Because the behavior of ideal gases, including Maxwell's exponential distribution of the kinetic energy of ideal gas particles, was what Boltzmann knew, it is also what Boltzmann used to illustrate his new methods.

The first of Boltzmann's calculations, appearing in a section entitled "Kinetic Energy Has Discrete Values," concerns the distribution of a given kinetic energy among a given number of identical systems, which he calls "molecules," when that energy can be divided into a finite number of identical pieces. Because Planck could have exploited Boltzmann's first calculation, we outline its steps (more succinctly than did Boltzmann). In doing so we adopt Boltzmann's notation in order that the results at the end of this section might be readily compared with his.

Consider, then, the problem of distributing energy L among n molecules. In order to count the number of partitions of this energy, Boltzmann divided the energy L into $\lambda\,[=L/\varepsilon]$ identical elements each of size ε. These elements of energy are distributed in such a way that ω_i of the molecules have energy $i\varepsilon$ where $i = 0,1,2,\ldots$.[10] Boltzmann's contention, one that has since 1877 borne fruit, is that a state is characterized by the *distribution* or set of numbers $\{\omega_0, \omega_1, \omega_2, \ldots\}$ and that the distribution with the largest number of ways P may be achieved subject to constraints on the system, that is, in this case, subject to a constant number of molecules n and a constant energy L, describes the state of thermodynamic equilibrium.

The number of ways this energy can be distributed over this number of molecules is "n choose $\{\omega_0, \omega_1, \omega_2, \ldots\}$," that is,

$$P = \frac{n!}{\omega_0!\,\omega_1!\,\omega_2!\ldots} \tag{9.1}$$

where P denotes the number of *permutations*, n is the number of molecules, ω_i is the number of molecules with energy $i\varepsilon$, and $\lambda[=L/\varepsilon]$ is the number of identical pieces or *quanta* of energy. Thus, the distribution $\{\omega_0, \omega_1, \omega_2, \ldots\}$ corresponding to the state of thermodynamic equilibrium is one that maximizes P, as given in (9.1), subject to the constraints

$$n = \sum_{i=0}^{\infty} \omega_i \tag{9.2}$$

and

$$\lambda = \sum_{i=0}^{\infty} i\omega_i \tag{9.3}$$

where n and λ are constants.

Boltzmann found that the task of finding the distribution $\{\omega_0, \omega_1, \omega_2, \ldots\}$ that maximizes P is made easier by finding the maximum of $\ln P$ rather than of P. The logarithm of both sides of (9.1) is given by

$$\ln P = \ln n! - \sum_{i=0}^{\infty} \ln \omega_i!. \tag{9.4}$$

In adopting Stirling's approximation, valid when $n >> 1$ and $\omega_i >> 1$ for all i, Boltzmann replaced $\ln \omega_i!$ with $\omega_i \ln \omega_i - \omega_i$ and $\ln n!$ with $n \ln n - n$. In this way he allowed the integers ω_i and n to be approximated with real numbers. Subsequently, (9.4) becomes

$$\begin{aligned} \ln P &= n \ln n - n - \sum_{i=0}^{\infty} \omega_i \ln \omega_i + \sum_{i=0}^{\infty} \omega_i \\ &= n \ln n - \sum_{i=0}^{\infty} \omega_i \ln \omega_i \end{aligned} \tag{9.5}$$

where $n \ln n$ is, of course, constant with respect to variations in the set $\{\omega_0, \omega_1, \omega_2, \ldots\}$ that makes $\ln P$, or equivalently $-\sum_{i=0}^{\infty} \omega_i \ln \omega_i$, stationary. Therefore, the set $\{\omega_0, \omega_1, \omega_2, \ldots\}$ that maximizes $\ln P$ is a solution of

$$\frac{\partial}{\partial \omega_j}\left[-\sum_{i=0}^{\infty} \omega_i \ln \omega_i - \alpha\left(\sum_{i=0}^{\infty} \omega_i - n\right) - \beta\left(\sum_{i=0}^{\infty} i\omega_i - \lambda\right)\right] = 0 \tag{9.6}$$

where α and β are Lagrange multipliers. This leads to

$$-\ln \omega_j - \alpha - \beta j = 0, \tag{9.7}$$

that is, to

$$\omega_i = e^{-\alpha} e^{-\beta i} \tag{9.8}$$

where in (9.8) we have replaced index j with index i. Inspecting the second derivative of the quantity in the square brackets of (9.6) for ω_i verifies that solution (9.8) maximizes $\ln P$, or equivalently, maximizes $-\sum_{i=0}^{\infty} \omega_i \ln \omega_i$.

Our next task is to use constraints (9.2) and (9.3) to eliminate the Lagrange multipliers α and β. Applying constraint (9.2) to result (9.8) we find that

$$\begin{aligned} n &= e^{-\alpha} \sum_{i=0}^{\infty} e^{-\beta i} \\ &= \frac{e^{-\alpha}}{1-e^{-\beta}}. \end{aligned} \tag{9.9}$$

Therefore,

$$\frac{\omega_i}{n} = \left(1-e^{-\beta}\right) e^{-\beta i}. \tag{9.10}$$

Applying constraint (9.3) to result (9.10) produces

$$\frac{\lambda}{n} = \left(1-e^{-\beta}\right) \sum_{i=0}^{\infty} i e^{-i\beta}. \tag{9.11}$$

Completing the sum in (9.11), with the aid of the identity

$$\sum_{i=0}^{\infty} i x^i = \frac{x}{(1-x)^2}, \tag{9.12}$$

which is valid when $x < 1$, yields

$$\frac{\lambda}{n} = \frac{1}{e^{\beta} - 1} \tag{9.13}$$

which, in turns, implies that

$$\beta = \ln\left(1+\frac{n}{\lambda}\right). \tag{9.14}$$

Equation (9.14) turns solution (9.10) into

$$\frac{\omega_i}{n} = \frac{n/\lambda}{(1+n/\lambda)^{i+1}} \tag{9.15}$$

where $i = 0,1,2\ldots$. This result is exact, given the suppositions of Boltzmann's model, and is identical to a corresponding expression derived by Boltzmann.

We note that this result (9.15) in no way limits the partition to that of *kinetic* energy. Rather (9.15) describes the distribution of λ *elements* (Planck's word) of any kind of energy among n identical molecules that are in thermal equilibrium with one another.

Boltzmann gave neither this nor any other interpretation of (9.15). Rather, this calculation and this result seem, in his 1877 paper, to be a mere illustration of the mathematics of his statistical theory of thermal equilibrium—a mathematics that he eventually applied to a more convincing model of an ideal gas. However, (9.15) was perfectly suited for Planck's problem in 1900–1901.

9.4 A Continuation of Boltzmann's First Calculation

Just as Boltzmann applied his newly developed statistical methods to a system, the ideal gas, already well known to him, Planck applied Boltzmann's methods to a system, blackbody radiation, about which, from his zeroth derivation, he knew quite a lot. What follows here is the outline of a calculation Planck *could have used* to derive the spectral energy density of blackbody radiation, had he carried Boltzmann's first calculation a few steps further. This continuation of Boltzmann's derivation consists chiefly of substituting the maximizing distribution (9.15) into $\ln P$ expressed in (9.4) and identifying the latter as proportional to the entropy. Planck's actual route to his desired result, his "first derivation," took another form—one that we present in chapter 10.

Boltzmann had calculated, in equation (9.15), the number ω_i, where $i = 1,2,\ldots\infty$, of n identical subsystems or molecules each having energy $i\varepsilon$ and sharing a total energy L. Recall that, in Boltzmann's usage, λ is the total number of identical elements of energy into which L is divided.

In the following calculation we transition from Boltzmann's notation to one closer to Planck's. Accordingly, we divide the total system energy L by the number n of identical molecules to get the average energy U_1 of one molecule. Therefore,

$$U_1 = \frac{L}{n}. \tag{9.16}$$

Furthermore, if we divide the total system energy L by the total number λ of identical energy elements we get the energy of one such element, which we (and Boltzmann) denote ε. Therefore,

$$\varepsilon = \frac{L}{\lambda}. \tag{9.17}$$

Dividing ε by U_1 we get

$$\frac{\varepsilon}{U_1} = \frac{n}{\lambda}. \tag{9.18}$$

Using (9.18) to replace n/λ in Boltzmann's equation (9.15) with its equivalent ε/U_1 produces

$$\frac{\omega_i}{n} = \frac{\varepsilon/U_1}{(1+\varepsilon/U_1)^{i+1}}, \tag{9.19}$$

that is, produces, in Planck's notation, the result of Boltzmann's first calculation.

From Boltzmann's calculation (9.5) of $\ln P$ we know that

$$\begin{aligned}\frac{\ln P}{n} &= \ln n - \sum_{i=0}^{\infty}\left(\frac{\omega_i}{n}\right)\ln \omega_i \\ &= -\sum_{i=0}^{\infty}\left(\frac{\omega_i}{n}\right)\ln\left(\frac{\omega_i}{n}\right).\end{aligned} \tag{9.20}$$

Using (9.19) we find that (9.20) is equivalent to

$$\begin{aligned}\frac{\ln P}{n} = \ln n - \left(\frac{\varepsilon}{U_1}\right) x \Bigg[\left\{\ln\left(\frac{n\varepsilon}{U_1}\right) - \ln\left(1+\frac{\varepsilon}{U_1}\right)\right\}\sum_{i=0}^{\infty} x^i \\ - \ln\left(1+\frac{\varepsilon}{U_i}\right)\sum_{i=0}^{\infty} i x^i\Bigg]\end{aligned} \tag{9.21}$$

where $x = (1+\varepsilon/U_1)^{-1}$. The sums contained in (9.21) are standard geometrical sums that can be completed with the aid of (9.12) and the identity

$$\sum_{i=0}^{\infty} x^i = \frac{1}{1-x}, \tag{9.22}$$

both of which are valid as long as $x < 1$. Using (9.12) and (9.22) to complete (9.21) produces, after some algebra,

$$\begin{aligned}\frac{\ln P}{n} &= \ln n - \left[\ln\left(\frac{n\varepsilon}{U_1}\right) - \ln\left(1+\frac{\varepsilon}{U_1}\right) - \left(\frac{U_1}{\varepsilon}\right)\ln\left(1+\frac{\varepsilon}{U_1}\right)\right] \\ &= -\left(\frac{U_1}{\varepsilon}\right)\ln\left(\frac{U_1}{\varepsilon}\right) + \left(1+\frac{U_1}{\varepsilon}\right)\ln\left(1+\frac{U_1}{\varepsilon}\right).\end{aligned} \tag{9.23}$$

Therefore, given Boltzmann's insight that $S \propto \ln P$, we find that these calculations produce

$$\frac{S}{n} \propto -\left(\frac{U_1}{\varepsilon}\right)\ln\left(\frac{U_1}{\varepsilon}\right) + \left(1+\frac{U_1}{\varepsilon}\right)\ln\left(1+\frac{U_1}{\varepsilon}\right). \tag{9.24}$$

The left-hand side of (9.24), S/n, is the average entropy S_1 of a single subsystem, while the right-hand side is a function of the average energy U_1 of a single subsystem. Note that the entropy S of the entire n-molecule system is, indeed, an extensive function of its total energy nU_1.

The parameter ε represents an element of energy that, apparently, was of little interest to Boltzmann, for he does not comment upon it. However, ε is quite interesting to us. For (9.24), with a constant k that turns the proportionality into an equality, provides an expression for the entropy function $S_1(U_1)$ that Planck needed in order to produce the function $U_1(T)$ via the equation of state $T^{-1} = \partial S_1/\partial U_1$. (See section 10.3 for a verification of this statement.) Planck's expression for $S_1(U_1)$, the fundamental relation (7.33), and Wien's displacement law (5.26) complete the functional dependence of the spectral energy density $u_\nu(T,\nu)$ of blackbody radiation (8.27) that Planck's zeroth derivation had already produced.

Could Planck have worked through this conceptually obvious, if algebraically complex, derivation to produce his desired result? Certainly! For we know that Planck was familiar with Boltzmann's 1877 paper on the statistical basis of entropy.[11] Did he actually work through these details? We do not know. In any case, Planck found another route to (9.24)—a route that is part of what we now call Planck's "first derivation."

9.5 The Boltzmann Factor

In addition to Planck's zeroth (chapter 8) and first derivations (chapter 10) of the spectral energy density of blackbody radiation and the derivation of section 9.4 by which Planck could have produced the expression $S_1(U_1)$, two other routes to blackbody radiation were forged by Albert Einstein, one in 1907 and another in 1917. These are taken up in chapters 11 and 12.

Einstein's two approaches to blackbody radiation depend upon a basic result of Boltzmann's statistical mechanics that is variously called the *Boltzmann factor*, the *Boltzmann distribution*, and the *canonical distribution*. The idea is both simple and profound and has many applications to both classical and quantized systems.

Boltzmann's factor concerns a system that could be a Planck resonator or a molecule embedded in a crystal or one moving freely in a gas—or even groups of such Planck resonators or molecules in various situations. The only requirement imposed on the system is that it exchanges energy and is in thermal equilibrium with a much larger system variously called the *environment* or the *temperature reservoir*. The purpose of the environment or reservoir is to fix the temperature of the smaller system.

The system of interest and the reservoir, with which it is in thermal contact, share a constant total energy E of which the system has share E_s and the reservoir has the remaining part $E-E_s$. Thus, the more energy contained in the system of interest, the less energy contained in the reservoir. Let $P(E_s)$ be the probability of a single microstate of the system when the energy of that microstate is E_s. This probability is directly proportional to the number of reservoir microstates with energy $E-E_s$, that is,

$$\begin{aligned} P(E_s) &\propto \Omega_r(E-E_s) \\ &\propto \exp[S_r(E-E_s)/k] \end{aligned} \tag{9.25}$$

where $\Omega_r(E-E_s)$ is the multiplicity of the reservoir, $S_r(E-E_s)$ is the entropy of the reservoir, and these are linked by $S_r(E-E_s)=k\ln\Omega_r(E-E_s)$. Given that the reservoir is much larger than the system so that $E \gg E_s$, we may expand the reservoir entropy through its first two terms in E_s so that

$$S_r(E - E_s) \approx S_r(E) - E_s \frac{dS_r(E)}{dE} \\ \approx S_r(E) - \frac{E_s}{T}. \tag{9.26}$$

Using (9.26) in (9.25) yields

$$P(E_s) \propto e^{S_r(E)/k} e^{-E_s/kT} \tag{9.27}$$

where $e^{-E_s/kT}$ is called the *Boltzmann factor.* Equation (9.27) expresses the probability $P(E_s)$ that a system occupies a single microstate having energy E_s when in thermal equilibrium with a much larger reservoir with temperature T. This probability decreases exponentially with system energy E_s.

Note that the factor $e^{S_r(E)/k}$ in (9.27) is the same for all microstates of the system, that is, for all values of the system energy E_s. Therefore, the ratio of the probabilities of two different microstates of the system having two different energies, E_1 and E_2, is

$$\frac{P(E_1)}{P(E_2)} = \frac{e^{-E_1/kT}}{e^{-E_2/kT}}. \tag{9.28}$$

Alternatively, the probability $P(E)$ of a system having energy E when in contact with a thermal reservoir with temperature T may be expressed as

$$P(E) = Ce^{-E/kT} \tag{9.29}$$

where the proportionality constant C is determined by summing both sides of (9.29) over all possible microstates i so that $C\sum_i e^{-E_i/kT} = 1$. The result,

$$P(E_j) = \frac{e^{-E_j/kT}}{\sum_i e^{-E_i/kT}}, \tag{9.30}$$

produces the probability of microstate *j* where the sum in the denominator is over all possible microstates *i*. These expressions, (9.28), (9.29), and (9.30), are important because of their many applications—including those related to the specific heat of a crystalline solid and to the quantum theory of radiation discovered, respectively, by Einstein in 1907 and 1917.

10 Planck's "First Derivation," 1900–1901

10.1 The "First Derivation"

The blackbody radiation law that Planck proposed on October 19, 1900 has stood the test of time.[1] And because Planck's law conformed to the available data, he immediately set to work searching for a physically grounded derivation. He later referred to this search as "the most strenuous work of my life."[2] By December 14 he had his derivation,[3] and by January he had elaborated it and submitted it to the editors of the *Annalen der Physik*.[4] The derivation presented in December and January, considered as a single argument, is called the "first derivation."

The derivation has two outstanding features. The first is Planck's adoption of Boltzmann's probabilistic relation for the entropy,

$$S \propto \log P \tag{10.1}$$

where the *permutability* P is a function of energy U. Interestingly, Boltzmann never identified the proportionality constant k that turns $S \propto \log P$ into $S = k \log P$, a constant that is now called *Boltzmann's constant.* Rather, Boltzmann's purposes were satisfied with a mere proportionality between S and $\log P$, probably because, as Planck later remarked, "he never believed it would be possible to determine this [proportionality] constant accurately."[5] Planck himself chose the letter k after the German word *Konstante*. For a time k was called "Planck's constant."[6]

The second outstanding feature of the first derivation is usually portrayed as Planck's quantization of the resonator energy. But Planck did not characterize his mathematics in this way and another interpretation is likely. Planck knew of Ludwig Boltzmann's 1877 paper on the statistical or probabilistic interpretation of entropy. In much the same way that Boltzmann discretized the energy of an ideal gas, that is, as a mere convenient and illustrative device, Planck discretized the energy of his resonators as a mere convenient and illustrative device.[7] Then Planck invoked Wien's displacement law. And it was Planck's invocation of this classical requirement that established energy quanta as a permanent feature of Planck's radiation law. In this way, Planck saw these energy quanta as required by, rather than at odds with, the physics with which he was familiar—physics that today we call "classical."[8]

10.2 Planck's Program

Planck's first task was to determine an expression for the average energy U_1 of a single resonator with natural frequency ν in terms of the temperature T of the blackbody radiation with which the resonator is in equilibrium. Then, his fundamental relation,

$$u_\nu(T,\nu) = \frac{8\pi\nu^2}{c^2} U_1(T,\nu), \tag{10.2}$$

would determine the spectral energy density function $u_\nu(T,\nu)$ when the blackbody to which it applied is composed of oscillators with a range of natural frequencies. The spectral energy density so determined must, of course, be consistent with Wien's displacement law

$$u_\nu(T,\nu) = \nu^3 \Theta\left(\frac{T}{\nu}\right) \tag{10.3}$$

where the function $\Theta(T/\nu)$ is unspecified.

Planck's chosen thermodynamic system consisted of N identical resonators each having natural frequency ν and sharing a total energy U_{tot}. According to the additivity of thermodynamic energy, the average energy of one resonator U_1 is related to the total energy U_{tot} of N identical resonators by

$$U_1 = \frac{U_{tot}}{N}. \tag{10.4}$$

Furthermore, the function $U_{tot}(T,\nu)$ is related to the system temperature T by

$$\frac{1}{T} = \frac{\partial S_{tot}}{\partial U_{tot}} \tag{10.5}$$

where

$$S_{tot} = k \ln \Omega \tag{10.6}$$

in which we have replaced Boltzmann's permutability P with the synonymous *multiplicity* Ω, in part because, in his *Annalen* paper, Planck employed the symbol P for another purpose. The multiplicity Ω is, like the permutability, the number of ways a total energy U_{tot} may be distributed among the N identical resonators of the system given that the energy is divided into identical discrete elements. Note that because S_{tot} is, as we shall see, a function of only the thermodynamic variable U_{tot}, the partial derivative $\partial S_{tot}/\partial U_{tot}$ in (10.5) is equivalent to the total derivative dS_{tot}/dU_{tot}.

However, Planck's December 14, 1900 paper refers to resonators with multiple natural frequencies $\nu', \nu'', \nu''', \ldots$.[9] Indeed, a blackbody must be receptive of all electromagnetic waves that fall upon it. Therefore, Thomas Kuhn was right to argue that Planck's 1901 *Annalen* derivation is a part of a larger, more comprehensive analysis that, according to the December 14 paper (which itself is fragmentary), includes resonators of different kinds. Not until Planck's publication of the first book-length edition of his *Lectures*[10] (in 1906) did he compose all these pieces into a coherent whole—or so Kuhn argues.[11]

On the other hand, Planck seemed, in the *Annalen* paper, to have found a short route to his desired result. It was in following this route that Planck recognized and solved an important problem in applying $S = k\log\Omega$. For, if the energy U_{tot} of the N resonators of frequency ν could be subdivided indefinitely, the multiplicity Ω of the N-resonator system would diverge to infinity. In order to keep Ω finite yet still proportional, in some way, to the number of the system's different possible realizations, Planck imagined that the energy U_{tot} was subdivided into "P elements of size ε" so that[12]

$$P = \frac{U_{tot}}{\varepsilon}. \tag{10.7}$$

In this way, Planck adapted to his own problem not only Boltzmann's entropy, but also Boltzmann's example of determining the multiplicity of a system by subdividing its total energy into identical, discrete, energy elements.

Planck may also have recognized the need for these energy elements by working backward from the desired result of his zeroth derivation of October 19, 1900. We know, for instance, that earlier, by his own admission, Planck had worked backward from the Wien distribution to find the entropy function it implied.[13] That he worked backward from the complete radiation law to the entropy function it implied and then discerned that energy elements were necessary also seems possible.[14]

10.3 Boltzmann's Entropy and Planck's Combinatorics

Planck distributed the P energy elements of (10.7) over N identical, yet distinguishable, resonators with natural frequency ν in $(P+N-1)!/P!(N-1)!$ ways according to a combinatoric formula that we motivate in the following way. Imagine that P white balls represent the P energy elements to be distributed among N identical resonators and that $N-1$ vertical lines represent dividers that separate successive resonators to which the energy elements have been assigned—as illustrated in Figure 10.1 for $P=7$ and $N=5$.

These $P+N-1$ objects can be arranged in $(P+N-1)!$ ways. And, since both the energy elements and the dividers are separately identical and indistinguishable, only $(P+N-1)!/P!(N-1)!$ of the $(P+N-1)!$ ways are distinguishable from one another. Therefore,

$$\Omega = \frac{(P+N-1)!}{P!(N-1)!} \tag{10.8}$$

expresses the multiplicity of the system.[15] The entropy S_{tot} associated with the N resonators is then given by

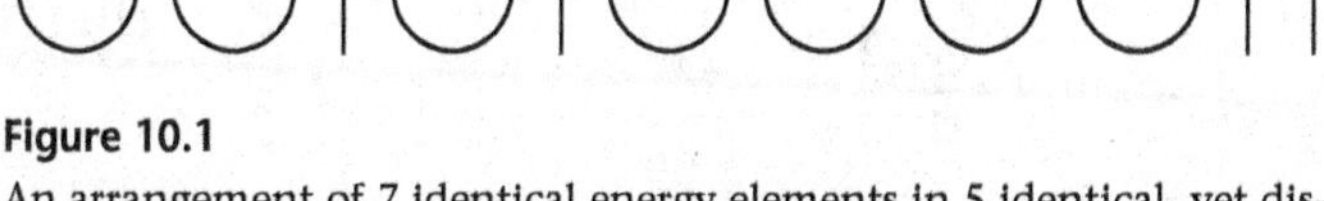

Figure 10.1
An arrangement of 7 identical energy elements in 5 identical, yet distinguishable (that is, ordered), resonators. The symbol O stands for an element of energy and the symbol l stands for a divider between different resonators. From left to right the five resonators possess, respectively, 2, 1, 4, 0, and 0 energy elements.

$$S_{tot} = k\ln\left[\frac{(P+N-1)!}{P!(N-1)!}\right] \tag{10.9}$$

and so

$$\begin{aligned} S_{tot} &= k[(P+N)\ln(P+N) - P\ln P - N\ln N] \\ &= kN\left[\left(1+\frac{P}{N}\right)\ln\left(1+\frac{P}{N}\right) - \frac{P}{N}\ln\frac{P}{N}\right] \\ &= kN\left[\left(1+\frac{U_{tot}}{N\varepsilon}\right)\ln\left(1+\frac{U_{tot}}{N\varepsilon}\right) - \frac{U_{tot}}{N\varepsilon}\ln\frac{U_{tot}}{N\varepsilon}\right]. \end{aligned} \tag{10.10}$$

The first step of equation (10.10) follows from applying Stirling's approximation, $\ln n! \approx n\ln n - n$ when $n >> 1$, to (10.9). In the present case, $P >> 1$ and $N >> 1$. The second step resorts to the definition, $P = U_{tot}/\varepsilon$, of the number of energy elements. Note that, although we (and Planck) call the N subsystems "resonators," their only property, as yet, is that they possess energy.

Note that, according to (10.10), the entropy S_{tot} of N resonators is a function $S_{tot}(U_{tot})$ of the energy U_{tot} of N resonators in such a way that increasing (or decreasing) N by a factor changes the total entropy by that same factor. This means that the entropy S_{tot} is, indeed, an extensive function of its energy U_{tot}—a condition of which, as argued in section 8.2, Planck seemed not aware.

Taking the energy derivative of the entropy dS_{tot}/dU_{tot} $[=1/T]$ we find that

$$\begin{aligned} \frac{1}{kT} &= \frac{1}{\varepsilon}\left[\ln\left(1+\frac{U_{tot}}{N\varepsilon}\right) - \ln\left(\frac{U_{tot}}{N\varepsilon}\right)\right] \\ &= \frac{1}{\varepsilon}\ln\left(1+\frac{N\varepsilon}{U_{tot}}\right) \end{aligned} \tag{10.11}$$

and, therefore, that

$$U_{tot} = \frac{N\varepsilon}{e^{\varepsilon/kT}-1}, \tag{10.12}$$

or, given $U_{tot} = NU_1$, that

$$U_1 = \frac{\varepsilon}{e^{\varepsilon/kT} - 1}. \tag{10.13}$$

Thus, relation (10.13), connecting the average internal energy U_1 of a single resonator to the temperature T of the system of resonators with which it is in equilibrium, follows from the relation $S = k \ln \Omega$ and the supposition that the total resonator energy is subdivided into energy elements of size ε.

10.4 The Program Completed

The work of completing Planck's derivation of the spectral energy density $u_\nu(T,\nu)$ and endowing blackbody radiation with electromagnetic properties falls to Planck's purely classical fundamental relation

$$u_\nu(T,\nu) = \left(\frac{8\pi\nu^2}{c^3}\right) U_1(T,\nu) \tag{10.14}$$

and to Wien's classical displacement law

$$u_\nu(T,\nu) = \nu^3 \Theta\left(\frac{T}{\nu}\right). \tag{10.15}$$

Equations (10.13), (10.14), and (10.15) together require that $\varepsilon = h\nu$ where h is a proportionality constant that Planck called "the quantum of action" and together determine the function $\Theta(T/\nu)$ $[= (8\pi h/c^3)(e^{h\nu/kT} - 1)^{-1}]$. Planck's ordering of the steps in his derivation suggests that the identification $\varepsilon = h\nu$ was for Planck a requirement of, rather than a departure from, classical physics. In his words, "The fact that the chosen energy element ε for a given group of resonators must be proportional to frequency ν follows immediately from the extremely important so-called Wien displacement law."[16] In this way Planck derived the spectral energy density

$$u_\nu(T,\nu)=\frac{8\pi\nu^2}{c^3}\frac{h\nu}{e^{h\nu/kT}-1} \tag{10.16}$$

in terms of two new fundamental constants h and k whose values he estimated by fitting (10.16) to the available data on blackbody radiation.

10.5 Planck's Natural Units

Planck's spectral energy density function (10.16) introduced two new fundamental constants now known, respectively, as Planck's constant h and Boltzmann's constant k. Planck was impressed that these fundamental constants together with the speed of light c and the constant of universal gravitation G could be combined in ways to define a fundamental mass m_o, length l_o, time t_o, and temperature T_o. In particular, one finds that

$$m_o=\sqrt{\frac{ch}{G}}\;\left[=2.18\cdot 10^{-8}kg\right], \tag{10.17a}$$

$$l_o=\sqrt{\frac{Gh}{c^3}}\;\left[=1.62\cdot 10^{-35}m\right], \tag{10.17b}$$

$$t_o=\sqrt{\frac{Gh}{c^5}}\;\left[=5.39\cdot 10^{-44}s\right], \tag{10.17c}$$

and

$$T_o=\sqrt{\frac{hc^5}{Gk^2}}\;\left[=3.55\cdot 10^{32}K\right] \tag{10.17d}$$

where here we have used current values of these fundamental constants in order to compute their numerical values in SI units. Planck called these combinations *natural units* and directed attention to them eight times in his published works between 1899 and 1923.[17] We now call these combinations *Planck units*. Today other systems of units can be composed from other sets of fundamental constants.[18]

10.6 The Status of "Energy Elements"

Planck found it necessary to divide the total energy of his system into countable energy elements in order to derive his radiation law. In doing so he is said to have departed from classical continuity. And it is this departure that is often celebrated as the beginning of the quantum era.

However, as noted, Planck may have conceived this step as one equivalent to the one taken by Boltzmann in discretizing the energy of an ideal gas and as a requirement of rather than a departure from classical physics. If so, one feature of this discretization continued to puzzle Planck. For unlike the artifacts of Boltzmann's discretization, the proportionality between the energy element ε and the frequency ν, that is, the "quantum of action" h, remained in observable quantities. Kuhn argues that initially Planck saw this puzzling persistence of h as simply a reason for more investigation.

At what point Planck began to conceive of his potentially disposable "energy elements" as enduring "energy quanta" is a matter of debate. According to Kuhn it took Planck a number of years that extended through 1909.[19] Other historians of science argue that, on the basis of his correspondence, Planck began to realize that finite energy quanta are necessary to blackbody theory as early as 1905.[20] Ultimately, Planck became convinced that these energy elements were more than mere "mathematical juggling" and "play a fundamental role in physics."[21]

11 Einstein's Response, 1905–1907

11.1 Einstein's Initial Response to Planck's Quantum

While many found Planck's use of Boltzmann's statistical mechanics noteworthy, most early readers of Planck's work paid little attention to his "elementary quantum of action" h, noting merely that it produced an empirically satisfying law.[1] However, there were a few, including H. A. Lorentz (1853–1928), Paul Ehrenfest (1880–1933), and Albert Einstein (1879–1955), who by 1905 had noticed that Planck's discretization of the resonator energy into units of $h\nu$ was crucial to his derivation of an empirically successful expression for the spectral energy density of blackbody radiation.

Einstein in particular turned Planck's discretization of energy into a tool with which to investigate other phenomena. In this way Einstein gave Planck's discrete energy elements a life beyond blackbody radiation. However, according to Thomas Kuhn, Planck "did not publically acknowledge the need for discontinuity until 1909, and there is no evidence that he had recognized it until the year before."[2] It is Einstein's earlier response to Planck's radiation law, in the period 1905–1907, to which we now turn.

11.2 The Entropy of Blackbody Radiation in the Wien Limit

Einstein first noticed a similarity between the entropy of an ideal gas composed of point particles and the entropy of high-frequency electromagnetic waves selected from the spectrum of blackbody radiation. Exposing this similarity, which suggested that high-frequency radiation behaved like the particles of an ideal gas, composed the bulk of Einstein's 1905 paper "Concerning an Heuristic Point of View toward the Emission and Transformation of Light"[3]—a paper that is primarily known for its explanation of the photoelectric effect.

Recall that the entropy $S(U,V)$ of an ideal gas is, as presented in equation (2.29), of the form

$$\begin{aligned} S(U,V) &= nR\ln V + g(U) \\ &= k\ln V^N + g(U) \end{aligned} \tag{11.1}$$

where n is the number of moles, the gas constant $R[=kN_A]$ is equal to Boltzmann's constant k times Avogadro's number N_A, $N[=nN_A]$ is the number of ideal gas particles, and $g(U)$ is a function of the system energy U and not of its volume V. The volume dependence of the entropy $k\ln V^N$ is in the form of Boltzmann's expression for the entropy of a system, that is, $S \propto \ln P$ or $S = k\ln P$ where P is the number of microscopic states that can be occupied by the system. Apparently, V^N is proportional to the number of microscopic states that can be occupied by the N particles of an ideal gas in volume V. This dependence and this understanding must have been well known to Einstein's readers.

Einstein applied standard thermodynamic methods to a system composed of high-frequency, monochromatic blackbody radiation with frequencies between ν and $\nu + d\nu$ when $h\nu/kT >> 1$. The spectral energy density is, according to Planck,

$$u_\nu(T,\nu) = \left(\frac{8\pi\nu^2}{c^2}\right)\frac{h\nu}{e^{h\nu/kT}-1}, \tag{11.2}$$

which, in the regime for which $h\nu/kT \gg 1$, reduces to the Wien distribution

$$u_\nu(T,\nu) = \left(\frac{8\pi\nu^2}{c^2}\right)h\nu e^{-h\nu/kT}. \tag{11.3}$$

One could construct such a system, for instance, by passing blackbody radiation through a filter which excludes all but radiation with frequencies between ν and $\nu + d\nu$ where $h\nu/kT \gg 1$.

We adopt the notation of chapter 5 for the purpose of representing the thermodynamic variables of Einstein's system of high-frequency blackbody radiation. Thus, its internal energy is $VU_\nu^{\nu+d\nu}$ where the monochromatic, cumulative energy density $U_\nu^{\nu+d\nu}$ is related to the spectral energy density u_ν by

$$\begin{aligned} U_\nu^{\nu+d\nu} &= \int_\nu^{\nu+d\nu} u_\nu d\nu \\ &= u_\nu d\nu. \end{aligned} \tag{11.4}$$

Therefore, $VU_\nu^{\nu+d\nu}$ and $Vu_\nu d\nu$ are equivalent expressions for the internal energy of this system.

The fundamental constraint for this system of monochromatic blackbody radiation is

$$d\left(VU_\nu^{\nu+d\nu}\right) = Td\left(S_\nu^{\nu+d\nu}\right) - \left(\frac{U_\nu^{\nu+d\nu}}{3}\right)dV \tag{11.5}$$

where $S_\nu^{\nu+d\nu}$ and $U_\nu^{\nu+d\nu}/3$ are the entropy and pressure of the system. Rearranging the terms of (11.5) we find that

$$d\left(S_\nu^{\nu+d\nu}\right) = \left(\frac{1}{T}\right)d\left(VU_\nu^{\nu+d\nu}\right) + \left(\frac{U_\nu^{\nu+d\nu}}{3T}\right)dV \tag{11.6}$$

from which follow the formal equations of state

$$\left[\frac{\partial\left(S_\nu^{\nu+d\nu}\right)}{\partial\left(VU_\nu^{\nu+d\nu}\right)}\right]_V=\frac{1}{T} \tag{11.7}$$

and

$$\left[\frac{\partial\left(S_\nu^{\nu+d\nu}\right)}{\partial V}\right]_{VU_\nu^{\nu+d\nu}}=\frac{U_\nu^{\nu+d\nu}}{3T}. \tag{11.8}$$

Solving the Wien distribution law (11.3) for the inverse temperature produces

$$\frac{1}{T}=\frac{k}{h\nu}\ln\left[h\nu\left(\frac{8\pi\nu^2}{c^2u_\nu}\right)\right] \tag{11.9}$$

by which we eliminate the temperature dependence of the formal equation of state (11.7). Therefore,

$$\begin{aligned}\left[\frac{\partial\left(S_\nu^{\nu+d\nu}\right)}{\partial\left(VU_\nu^{\nu+d\nu}\right)}\right]_V&=\frac{k}{h\nu}\ln\left[h\nu\left(\frac{8\pi\nu^2}{c^2u_\nu}\right)\right]\\&=\frac{k}{h\nu}\left\{\ln V+\ln\left[h\nu\left(\frac{8\pi\nu^2 d\nu}{c^2VU_\nu^{\nu+d\nu}}\right)\right]\right\}\end{aligned} \tag{11.10}$$

where we have used (11.4). Integrating (11.10) yields

$$S_\nu^{\nu+d\nu}=k\ln V^{VU_\nu^{\nu+d\nu}/h\nu}-\left(VU_\nu^{\nu+d\nu}/h\nu\right)\ln\left(VU_\nu^{\nu+d\nu}/e\right)+f(V) \tag{11.11}$$

where $f(V)$ is an undetermined function of volume V. Einstein could also have arrived at (11.11) by eliminating the temperature from the second formal equation of state (11.8). According to Einstein, it must be that $f(V)=0$, since the entropy $S_\nu^{\nu+d\nu}$ of blackbody radiation within this frequency interval must vanish when its energy $VU_\nu^{\nu+d\nu}$ vanishes. In this case, (11.11) reduces to

$$\begin{aligned} S_\nu^{\nu+d\nu} &= k\ln V^{VU_\nu^{\nu+d\nu}/h\nu} - \left(VU_\nu^{\nu+d\nu}/h\nu\right)\ln\left(VU_\nu^{\nu+d\nu}/e\right) \\ &= k\ln V^{VU_\nu^{\nu+d\nu}/h\nu} + j\left(VU_\nu^{\nu+d\nu}\right), \end{aligned} \tag{11.12}$$

where the function $j(VU_\nu^{\nu+d\nu})$ is used in order to cast (11.11) into a form (11.12) that is the same as that of (11.1) describing the entropy of an ideal gas.

Comparing the resulting volume dependence $k\ln V^{VU_\nu^{\nu+d\nu}/h\nu}$ of the entropy of high-frequency, monochromatic blackbody radiation with the volume dependence $k\ln V^N$ of the entropy of an ideal gas suggested to Einstein that the factor $VU_\nu^{\nu+d\nu}/h\nu$ plays a role in a system of high-frequency blackbody radiation similar to the number N of gas particles in an ideal gas. Therefore, according to Einstein, the quantity $VU_\nu^{\nu+d\nu}/h\nu$ "behaves thermodynamically as if it [the system] consisted of a number of independent energy quanta,"[4] each quantum of which is localized and contains energy $h\nu$. In this way Einstein separated the energy quanta from the process of their absorption in and emission from Planck resonators. But this was only the first of two related themes Einstein addressed in his 1905 paper.

11.3 The Photoelectric Effect

Einstein's second theme concerns his explanation of Philipp Lenard's (1862–1947) then-recent observations on the electrons produced by irradiating metals with ultraviolet light. Lenard had found that the kinetic energy of the electrons ejected from metallic surfaces increases linearly with the frequency of the irradiating light and is independent of its intensity—a finding that both Lenard and Einstein found impossible to explain on the basis of "the usual conception that the energy of light is continuously distributed over the space through which it propagates."[5]

Einstein's view, in this paper, was that an electron is emitted when a localized quantum of light having energy $h\nu$ penetrates the metal and delivers its energy to a single electron. Given that an electron loses a certain minimum energy W in exiting the metal surface, the maximum kinetic energy K of an ejected electron is related to the light frequency ν and the metal's *work function* W by

$$K = h\nu - W. \tag{11.13}$$

Lenard measured this maximum kinetic energy K of the ejected electrons by applying a retarding potential to the metal relative to surrounding grounded structures. When this retarding potential reached a value ΔV that just prevented the ejected electrons from leaving the metal surface, $e\Delta V = K$, and, consequently,

$$\Delta V = \frac{h}{e}\nu - \frac{W}{e}. \tag{11.14}$$

According to (11.14), the retarding potential ΔV is a linear function of the irradiating frequency ν "whose slope [$= h/e$] is independent of the nature of the emitting surface."[6] Indeed, Einstein found that this prediction was in (order-of-magnitude) agreement with Lenard's data.[7]

11.4 The Einstein Solid

In 1905 Einstein observed that the theory of light quanta is "remarkable insofar as it facilitates the understanding of a series of regularities."[8] One of the regularities that Einstein explored in 1907 was the specific heat of solids. Einstein conceived of his calculation as both a rederiving of the Planck radiation law and of deriving, for the first time, an expression for the specific heat of a crystalline solid at thermodynamic temperatures extending down to absolute zero.

Interestingly, Einstein's dual analysis depends upon its application in two very different, even divergent and contradictory, environments, which, unfortunately for modern readers of Einstein's paper, are not made explicit. Even so the two derivations share much of the same mathematics. However, in the first derivation (of the spectral energy density), one needs to imagine a blackbody which receives all the radiation that falls upon it and, consequently, is composed of resonators with a continuum of different resonant frequencies, while, in the second derivation (of the specific heat of a solid), one needs to imagine a solid composed of identical resonators each with the same resonant frequency.

In the first derivation, Einstein adopted Planck's "fundamental relation,"

$$u_\nu(T,\nu) = \left(\frac{8\pi\nu^2}{c^3}\right) U_1(T,\nu), \tag{11.15}$$

as following from resonator dynamics and Maxwell's electrodynamics. Recall that the left-hand side of (11.15) is the spectral energy density of blackbody radiation $u_\nu(T,\nu)$ as a function of frequency ν, while the factor $U_1(\nu,T)$ on the right-hand side is the average energy of a single Planck resonator with natural frequency ν. Then Einstein derives an expression for $U_1(\nu,T)$ and, in this way, determines not only the spectral energy density $u_\nu(T,\nu)$ but also the heat capacity C_V of the solid where

$$C_V = N\left(\frac{\partial U_1}{\partial T}\right)_V \tag{11.16}$$

and N is the number of particles in the solid.

What Einstein and others knew, in 1907, of the heat capacity of solids was the remarkable fact that in most cases the molar specific heats C_V/n of room temperature solids were all close to the same number $25J/K$, a fact known as the *law of Dulong and Petit* after the French physicists Pierre Louis Dulong and Alexis Thérèse Petit who discovered it in 1819. The law of Dulong and Petit is

$$\frac{C_V}{n} = 3R \tag{11.17}$$

where R is the fundamental gas constant and n is the number of moles.

If the molecules that occupy the lattice sites of a homogeneous solid absorb and store the energy of radiation only in quanta of size $h\nu$, as suggested by Einstein's analysis, then the average energy of a resonator $U_1(T,\nu)$ must reflect this discontinuity. For the purpose of deriving the function $U_1(T,\nu)$ Einstein adopted the *Boltzmann* or *canonical distribution* that expresses the probability $P(j\varepsilon)$ that a system having energy $j\varepsilon$ in equilibrium with many others is given by

$$P(j\varepsilon) = \frac{e^{-j\varepsilon/kT}}{\sum_{i=0}^{\infty} e^{-i\varepsilon/kT}} \tag{11.18}$$

where $\varepsilon = h\nu$. Therefore, the average energy of a resonator or lattice molecule that oscillates in one dimension is given by

$$U_1(T,\nu) = \frac{\sum_{i=0}^{\infty}(ih\nu)e^{-ih\nu/kT}}{\sum_{i=0}^{\infty} e^{-ih\nu/kT}}. \tag{11.19}$$

Geometrical sums like those in the numerator and denominator of (11.19) have been used before in sections 9.3 and 9.4. Accordingly, equation (11.19) becomes

$$U_1(T,\nu) = \frac{h\nu}{e^{h\nu/kT} - 1}. \tag{11.20}$$

Result (11.20) reproduces Planck's earlier calculation of the average energy of a resonator U_1—one that, given the fundamental relation $\mu_\nu = (8\pi\nu^2/c^3)U_1$, reproduces the spectral energy density of blackbody radiation just as Einstein claimed.

Result (11.20) also produces an expression for the internal energy $U_{tot}[= NU_1]$ of a solid composed of N identical molecules or oscillators each one of which oscillates in one dimension with frequency ν. Therefore, the internal energy of a solid composed of N identical oscillators, each one of which oscillates in three dimensions, is given by

$$\begin{aligned} U_{tot}(T,\nu) &= 3NU_1(T,\nu) \\ &= \frac{3Nh\nu}{e^{h\nu/kT} - 1}. \end{aligned} \tag{11.21}$$

Therefore, its heat capacity is

$$\begin{aligned} C_V &= \frac{dU_{tot}}{dT} \\ &= 3nR\frac{e^{h\nu/kT}(h\nu/kT)^2}{\left(e^{h\nu/kT} - 1\right)^2} \end{aligned} \tag{11.22}$$

and its molar specific heat capacity is

$$\frac{C_V}{n} = 3R\frac{e^{h\nu/kT}(h\nu/kT)^2}{\left(e^{h\nu/kT} - 1\right)^2}. \tag{11.23}$$

At high temperatures and low frequencies such that $h\nu/kT \ll 1$, equation (11.23) recovers the law of Dulong and Petit, $C_V/n = 3R$.

In 1907 scientists had become aware of molar specific heat capacities that drop below their Dulong-Petit value at low temperatures. Einstein found a good example of such data in

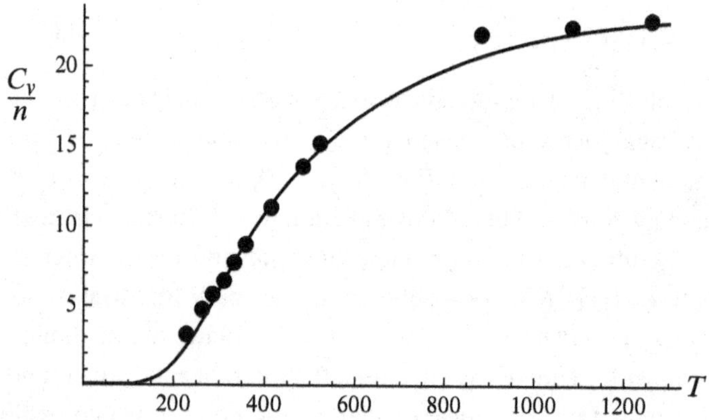

Figure 11.1.
Molar specific heat capacity C_V/n of diamond versus thermodynamic temperature T. Curve: Einstein's prediction (11.23) with $h\nu/k = 1325K$. Filled circles: data for diamond available to Einstein in 1907.

the molar specific heat capacity of diamond. We have drawn a graph of (11.23) in Figure 11.1 with an Einstein frequency $\nu[= k(1325K)/h]$ chosen in order to fit Einstein's data for diamond. As one can see, Einstein's theory fits the data fairly well—a success widely credited for initiating the modern study of solid-state physics.

12 Einstein on Emission and Absorption, 1917

12.1 Einstein's "Quantum Theory of Radiation"

As we have seen, Planck's derivation of the spectral energy density $u_\nu(T,\nu)$ of blackbody radiation extends over several complexly framed papers. In contrast, the derivation of $u_\nu(T,\nu)$ found in the first part of Einstein's 1917 paper "Zur Quantentheorie der Strahlung"[1] ("On the Quantum Theory of Radiation") consists of a handful of deceptively simple assumptions and a little algebra. While Planck's methods are largely forgotten, Einstein's helped develop the photon concept that underpins technologies such as the laser and maser. Undergraduate physics majors sometime read the first part of Einstein's 1917 paper. Even their teachers avoid reading Planck.

But not all contrasts are in favor of Einstein's derivation. For, while it is clear (to us) that Planck introduced the quantum by creating the energy elements shared among his resonators, discerning the entry of quantum physics into Einstein's derivation is not so easy.

Einstein's paper is divided into two parts. In the first part Einstein derives Planck's blackbody radiation formula from a few hypotheses concerning the emission and absorption of electromagnetic radiant energy from and by a system he calls a "molecule." In particular, Einstein requires that its energy is conserved. In the second part he shows that only when the emission and absorption of radiation are accomplished probabilistically by light quanta is momentum also conserved. Thus, light quanta are necessary in order that "we arrive at a consistent theory."[2]

However, the division of "On the Quantum Theory of Radiation" into two logically distinct parts presents Einstein's readers with a problem. For his energy argument seems to stand on its own apart from the momentum argument and, at the same time, seems to require no quantum hypothesis. This lack of a quantum hypothesis is especially apparent when viewing Einstein's derivation in the light of Planck's classical models of the emission and absorption of radiation as developed in chapter 7. And if the first part of Einstein's argument stands alone and requires no quantum hypothesis, then he has succeeded, where Planck had failed, in finding a classical derivation of the spectral energy density $u_\nu(T,\nu)$ as a function of frequency ν and temperature T.

Certainly this could not have been Einstein's understanding. His own statement that "Today we know that no consideration which is based on classical mechanics and electrodynamics can lead to a useful radiation formula" is unambiguous and, no doubt, correct. But in what way does quantum physics enter into the first part of Einstein's derivation? Here we seek an answer to this question.

12.2 Einstein's Derivation

Einstein, in his energy argument, postulates the following: [1] the spectral energy density $u_\nu(T,\nu)$ observes Wien's displacement law

$$u_\nu(T,\nu) = \nu^3\Theta\left(\frac{\nu}{T}\right) \tag{12.1}$$

where $\Theta(\nu/T)$ is an unknown function, [2] the absorbing and emitting molecule resides in certain states having discrete energies ε_n with $n = 1,2,\ldots$, [3] the probability W_n of the nth state is canonically distributed so that

$$W_n = p_n e^{-\varepsilon_n/kT} \tag{12.2}$$

where p_n is the (unknown) statistical weight of the state, [4] the probability that the molecule spontaneously emits electromagnetic energy of amount $\varepsilon_m - \varepsilon_n$ in transitioning from a higher energy state m to a lower one n in time dt is $A_m^n dt$, the probability that the spectral energy density u_ν stimulates the molecule to emit electromagnetic energy $\varepsilon_m - \varepsilon_n$ in time dt is $u_\nu B_m^n dt$, and the probability that the spectral energy density u_ν stimulates the molecule to absorb electromagnetic energy $\varepsilon_m - \varepsilon_n$ in time dt is $u_\nu B_n^m dt$, [5] in equilibrium, the rates of emission and absorption are equal so that

$$W_m\left(A_m^n + u_\nu B_m^n\right) = W_n u_\nu B_n^m, \tag{12.3}$$

and [6] in the high temperature limit as $T \to \infty$, the spectral energy density $u_\nu \to \infty$ and stimulated emission and absorption dominate the equilibrium relation (12.3) so that

$$p_m B_m^n = p_n B_n^m, \tag{12.4}$$

a relation called *microscopic reversibility* or *detail balance.*

Combining (12.2), (12.3), and (12.4) and solving for the spectral energy density u_ν produces

$$u_\nu(\nu,T) = \frac{A_m^n / B_m^n}{e^{(\varepsilon_m - \varepsilon_n)/kT} - 1}. \tag{12.5}$$

By requiring that this form (12.5) conform to Wien's displacement law (12.1), Einstein finds that

$$\frac{A_m^n}{B_m^n} = \alpha \nu^3 \tag{12.6}$$

where α is an undetermined combination of fundamental constants, and also that

$$\varepsilon_m - \varepsilon_n = h\nu \tag{12.7}$$

where h is Planck's constant. In 1913 Niels Bohr had adopted (12.7) as a postulate, which, in part, describes the emission and absorption spectra of the hydrogen atom. In Einstein's derivation Bohr's relation (12.7) is derived rather than assumed.

The result of (12.5)–(12.7) is the spectral energy density of blackbody radiation given by

$$u_\nu(\nu, T) = \frac{\alpha \nu^3}{e^{h\nu/kT} - 1}. \tag{12.8}$$

Note that, unlike the result of Planck's 1900–1901 derivation, Einstein's result (12.8) does not express the prefactor $\alpha[= 2h/c^2]$ in terms of fundamental constants.

12.3 Einstein's Derivation Made Classical?

Einstein states that postulate [2], invoking discrete energies ε_n, constitutes a quantum postulate. In his phrase, these states are "according to quantum theory." However, there seems to be no reason why the energies of Einstein's molecule must be discrete. Why should they not be distributed continuously as, say, are the translational energy states of a molecule of an ideal gas? In fact, Planck and others had established that the spectral energy density u_ν is a continuous function of ν for $\nu \geq 0$. Therefore, assuming continuously distributed molecular energies seems consistent with the desired spectral energy density (12.8).

A simple change in Einstein's argument allows us to assume that the energy of the molecule is continuously distributed.

Accordingly, we express the canonical probability that the molecule resides in energy state n with energy between ε_n and $\varepsilon_n + d\varepsilon_n$ as given by

$$dW_n = p(\varepsilon_n)e^{-\varepsilon_n/kT}d\varepsilon_n \tag{12.9}$$

where the probability W_n that the molecule resides in the nth state and the density-of-states function $p(\varepsilon_n)$ are now both continuous functions of the energy ε_n of the state.

Transitions between different energies, say between ε_m and ε_n, where these are now continuously distributed are still possible. In this case, the spontaneous and induced emission and absorption transition frequencies are continuous functions, respectively $A(m,n)$, $B(m,n)$, and $B(n,m)$, of the indices m and n. Energy balance is then

$$p(\varepsilon_m)e^{-\varepsilon_m/kT}\left[A(m,n)+u_\nu B(m,n)\right] = p(\varepsilon_n)e^{-\varepsilon_n/kT}u_\nu B(n,m). \tag{12.10}$$

If indefinitely high temperatures T produce indefinitely high spectral energy densities u_ν, then detail balance,

$$p(\varepsilon_m)B(m,n) = p(\varepsilon_n)B(n,m), \tag{12.11}$$

follows.

The steps of this argument unfold as before and produce Planck's spectral energy density as given in (12.8). Now, however, there is no doubt, because Einstein's molecule can occupy a continuum of energy states, that the spectral energy density $u_\nu(T,\nu)$ is a continuous function of the frequency ν.

12.4 Einstein's Missing Quantum Hypothesis

So, where does quantum physics enter Einstein's argument? There are several, manifestly classical inputs into Einstein's theory of quantum radiation: the Wien displacement law, the canonical distribution of energy, and detail balance. Furthermore, as we have shown, discretizing the molecule's energies is unnecessary in Einstein's derivation. Indeed, every one of Einstein's six postulates [1]–[6] has either a classical derivation or a classical analog. Induced emission and absorption, for instance, are classical processes. And, the equality between these rates plays a crucial role in deriving Planck's fundamental relation $u_\nu = (8\pi\nu^2/c^3)U_1$.

It has been asserted that "The sole quantum idea Einstein invoked was the concept of stationary states."[3] If "stationary states" means "nonradiating stationary states" as used by Bohr in his model of the hydrogen atom, and these states decay probabilistically, then yes, stationary states constitute a quantum hypothesis. But such was not Einstein's language. And both steady or stationary states and probabilistic events have classical analogs. For instance, a Planck resonator that receives the same energy as it radiates, as modeled in section 7.4, occupies a classical, stationary state.

There remains, from among postulates [1]–[6], spontaneous emission as a possible quantum postulate.[4] But spontaneous emission also has a classical analog, one that is integral to Planck's method of deriving his fundamental relation. After all, an isolated charged oscillator will, according to the principles of classical electromagnetism, radiate its energy, and this radiation will cause its oscillation amplitude to decay, as described in section 7.3.

Even so, Einstein used spontaneous emission in a way that never could have occurred to anyone as wedded to classical methods as Planck was. For the spontaneous emission of a Planck resonator is determined by its natural frequency ω_0 and by its decay rate $\sigma\omega_0^2/2$ $[= q^2\omega_0^2/12m\pi\varepsilon_0 c^3]$ and not by the difference between the energies of its initial and final states. It is

Einstein's unusual, nonclassical kind of spontaneous emission that marks his analysis rather than the bald fact of "spontaneous emission."

To summarize, Einstein like Planck before him depended upon the Wien displacement law (12.1) to introduce the radiation frequency ν into a form (12.5) not previously containing that frequency. And, since the Wien displacement law is a classical requirement, its important role drew Planck's attention away from the quantum physics he assumed. In hindsight, though, we easily locate the quantum postulate adopted by Planck in his quantization of the energy elements.

The quantum hypothesis in Einstein's 1917 theory of quantum radiation is not so easily located and may lie in the nonclassical behavior Einstein attributed to spontaneous emission. In any case, Einstein's 1917 paper on the quantum theory of radiation seems not to clearly identify a quantum postulate.

The Big Ideas

The form of our text is that of an extended mathematical argument. And attending to the mathematics sometimes distracts us from the big ideas guiding the argument. As a corrective, we verbally narrate these ideas here—especially those that we have found surprising.

James Clerk Maxwell's achievement in fashioning his eponymous equations stands above all others in making sense of our daily experience. For we have only to open our eyes to see the glory of the visible world that Maxwell's equations so precisely describe. Under the special condition that Maxwell's electromagnetic waves are in thermal equilibrium with the objects that produce and absorb them, they compose a thermodynamic system. It is this thermodynamic system that Planck sought to describe and to understand.

Of course, Planck built upon a foundation laid by others (chapter 1). In 1859 Gustav Kirchhoff conceived of blackbody radiation as a physical object to which one could apply the recently articulated laws of thermodynamics (chapters 2 and 3). But it was Ludwig Boltzmann who, in 1884, developed the idea that blackbody radiation is a fluid that can be assigned a definite volume, pressure, and temperature and that can do work or suffer work to be done upon it (chapter 4). Thus, blackbody radiation is a system to which the two laws of classical thermodynamics apply in much the same way as those laws apply to an ideal gas. Boltzmann also demonstrated that his thermodynamic description of blackbody radiation, unlike that of an ideal gas, has no adjustable parameters and applies in all temperature, pressure, and density regimes. It was this universality that attracted Planck, in 1894, to begin his theoretical investigation of blackbody radiation.

Just as Maxwell "looked under the hood" of the ideal gas and saw molecules moving around with speeds governed by a definite distribution, one we now call a "Maxwellian" distribution, Planck looked under the hood of blackbody radiation and saw electromagnetic waves propagating in all directions with frequencies governed by a definite distribution of intensity that had a name, the *spectral energy density*, but no known form. Planck's derivation of this form consists of combining four pieces of physics: Wilhelm Wien's displacement law (chapter 5), Max Planck's "fundamental relation" (chapter 7), Boltzmann's statistical or probabilistic definition of entropy (chapter 9), and Planck's combinatoric formula (chapter 10).

Wien's 1893 derivation of his displacement law, in turn, depends upon Boltzmann's 1884 thermodynamic derivation of the energy equation of state of blackbody radiation called the *Stefan-Boltzmann law* (chapter 4). Louis Buchholtz's translations of Boltzmann's 1884 and Wien's 1893 papers are found in Appendices A and B below. To our surprise, these derivations accomplish less than their reputations had led us to expect. In fact, Boltzmann's derivation produces an adiabatic invariant[1] that mimics the form of the Stefan-Boltzmann law (section 4.7), while Wien's derivation (section 5.9) also produces an adiabatic invariant and is in a form that is, at least superficially, different from the one with which we are familiar.

For instance, Boltzmann's 1884 derivation shows only that the combination u/T^4 is an adiabatic invariant. (Here u is the energy density of blackbody radiation and T is its absolute temperature.) According to our current understanding of the Stefan-Boltzmann law, the combination u/T^4 is an absolute rather than an adiabatic invariant. But Boltzmann's paper falls short of demonstrating this satisfying result.

Wien's derivation, as summarized in section 5.9, is similarly deficient. He combines different forms of the adiabatic invariant of blackbody radiation and a newly discovered electromagnetic adiabatic invariant to produce a result that is consequently an adiabatic invariant. While the adiabatic invariants Wien and Boltzmann derive are each a single step away from the well-known form of these laws (sections 2.10, 4.4, and 5.5), their unfinished state prompts the pedagogue within in us to admonish, "Close, but no cigar."

Also interesting, but perhaps not surprising, is that in these derivations Boltzmann and Wien pictured blackbody radiation as contained within a cylinder that was compressed or expanded with a slowly moving piston. In contrast, modern derivations, such as that of the Stefan-Boltzmann law in section 4.2 and that of Wien's displacement law in sections 5.3, 5.4, and 5.5, are entirely algebraic.

The "fundamental relation" is Planck's first significant contribution to solving the problem of blackbody radiation. Here Planck shows how a resonator (or damped, driven harmonic oscillator with natural frequency ν) can be in equilibrium with blackbody radiation at absolute temperature T. According to the fundamental relation, the spectral energy density of blackbody radiation $u_\nu(T,\nu)$ is directly proportional to the average energy $U_1(T,\nu)$ of a single resonator. We take care to motivate and articulate each step of Planck's derivation of the fundamental relation (chapters 6 and 7) while avoiding one of his circumlocutions by directly employing the Dirac delta function (not invented by Paul Dirac until 1930).

The fundamental relation allowed Planck to trade one problem for another, supposedly easier one, that is, to trade finding the spectral energy density $u_\nu(T,\nu)$ for finding the average energy $U_1(T,\nu)$ of one resonator when immersed in blackbody radiation. Given the function $U_1(T,\nu)$ Planck would need only to require that the spectral energy density $u_\nu(T,\nu)$ resulting from the fundamental relation observe the form required by Wien's displacement law.

Planck's initial attempt at finding $U_1(T,\nu)$, in March of 1900, involved some guesswork and produced only Wien's incomplete (and thus incorrect) spectral energy density of 1896 that we call *Wien's distribution.* Thomas Kuhn claimed that Planck's derivation was faulty, because the criteria Planck imposed on the spectral energy density function were not selective enough to produce a unique solution.[2] In particular, Planck ignored the possibility that the entropy of his system of resonators could be an extensive function of its parts (sections 8.2, 8.3, and 8.5). And since the entropy of Planck's resonators is, indeed, extensive, the basic equation upon which Planck founded the March 1900 derivation reduces to an uninformative tautology. Today no one would make this mistake. After all, extensivity is now recognized as a common feature of thermodynamic systems. But in Planck's time the meaning of extensivity and of a number of other thermodynamics concepts had yet not been fully digested. Even a thermodynamics expert like Planck[3] can make mistakes!

Planck's first successful attempt to construct the spectral energy density in October of 1900, which we call "Planck's zeroth derivation" (section 8.5), is sometimes demeaned as "curve fitting." Even Planck devalued it as "mere guesswork." His guesswork consisted of postulating a form for the second derivative of the entropy of a system of resonators with respect to resonator energy. When integrated twice this form produces the entropy as a function of resonator energy. Then Planck determined the desired function $U_1(T,\nu)$ from the thermodynamic relations between entropy, energy, and temperature and required the resulting spectral energy density $u_\nu(T,\nu)$ to conform to Wien's displacement law. While this procedure produces a result that has stood the test of time, Planck's postulate hides the physical basis of this derivation.

Planck's most convincing assault on the problem of blackbody radiation (sections 10.1 and 10.2), his "first derivation" (the first derivation in which a quantum hypothesis could be identified), adopted Boltzmann's probabilistic definition of the entropy of a thermodynamic system (chapters 9). In doing so Planck became the first to apply Boltzmann's method to a system other than the ideal gas. In particular, Planck took Boltzmann's definition of the entropy, $S \propto \log\Omega$, transformed it into $S = k\log\Omega$, and devised or discovered a combinatoric formula for determining the multiplicity Ω. This multiplicity Ω is, in this context, equivalent to what Boltzmann called the "permutability" P and what Planck later called the "probability" W.

Surprisingly, Planck could have adopted not only Boltzmann's 1877 definition of entropy but also Boltzmann's first calculation of the multiplicity Ω found in his 1877 paper. Here Boltzmann discretized the energy of an ideal gas particle by dividing it into a discrete number of identical energy elements of size ε that were then distributed over N particles. We know that Planck was familiar with this paper.[4] And while Boltzmann was concerned with ideal gases, this discretization could also have applied to Planck's resonators. Thus, Planck could have adapted Boltzmann's discretization to his own use—as was demonstrated in section 9.4.

We don't know why Planck did not pursue Boltzmann's ready-made route. Instead Planck devised or discovered his own way of determining the multiplicity of ways Ω in which a large number of energy elements of size ε could be distributed over a large number of resonators (section 10.3). Boltzmann's entropy and Planck's combinatorics together produced an expression for the average entropy of a single resonator in terms of its average energy. From this relation, the system temperature T and the function $U_1(T,\nu)$ follow. Then the fundamental relation determines the spectral energy density $u_\nu(T,\nu)$.

Requiring that $u_\nu(T,\nu)$ observe the form required by Wien's displacement law is Planck's last step (section 10.4). Interestingly, enforcing Wien's displacement law requires that Planck's energy element ε be directly proportional to resonator frequency ν, that is, requires that $\varepsilon = h\nu$ where the proportionality constant h soon came to be known as "Planck's constant." However, recall that, for Planck as well as for Boltzmann, the division of the total energy of a collection of resonators into elements of finite size ε was a device whose purpose was to make the system multiplicity Ω finite rather than infinite. For this reason, Planck missed the significance of the proportionality constant h. Consequently, he was perplexed when he found that he could not follow Boltzmann in making h vanishingly small in his final result. For such would render the spectral energy density inconsistent with the available experimental data.

We recapitulate the reasons why Planck missed the significance of the constant h. In Planck's zeroth derivation of

October 1900 h originates as one of two integration constants of a postulated second-order differential equation. And, in Planck's first derivation of 1900–1901 (chapter 10) and in Einstein's derivation of 1917 (chapter 12), Wien's (classical) displacement law requires that the size of the energy element ε be such that $\varepsilon \propto \nu$, that is, such that $\varepsilon = h\nu$ where h is constant. Thus, nowhere in this narrative does the appearance of Planck's constant h seem to be more than a classical device or requirement. Such was Planck's unwittingly introduction of a quantum hypothesis.

Historians of science and physicists alike are particularly interested in the origin of the constant h and its role in creating a new kind of nonclassical or quantum physics. But herein lies an irony. Today Planck's unwitting introduction of the "quantum of action" h is better known and better understood than the more technically demanding, classical parts of Planck's derivations, that is, better known than Wien's displacement law and Planck's fundamental relation. Our hope is that our book will help remedy this imbalance.

Chapter 12, our final chapter, addresses Einstein's 1917 paper "Quantum Theory of Radiation." In it Einstein presents a new, relatively simple, and now widely studied derivation of the spectral energy density of blackbody radiation. Our question is, "At what point did Einstein introduce a quantum hypothesis into this derivation?" To our surprise this question has no easy answer. Einstein's own remark that his discrete states "are according to quantum theory" is not pertinent, since, as we show (section 12.3), one can also derive Einstein's result even when the energy of Einstein's "molecule" is continuously distributed.

Acknowledgments

The authors gratefully acknowledge Robert C. Hilborn for providing copies of original sources, for entering into discussions related to the quantum presuppositions behind Einstein's 1917 paper on the quantum theory of radiation, for formulating a continuous version of Einstein's energy states (section 12.3), and for critically commenting on and improving our derivation of Wien's 1893 displacement law.

Each of the authors also wants to acknowledge their teachers for introducing them to the wonderful world of physics and for personally modeling excellence, kindness, and character: Don Lemons (Robert L. Armstrong and S. Peter Gary), William Shanahan (John W. Priess and Sam B. Treiman), and Louis Buchholtz (Robert Weinstock, Alexander L. Fetter, and Dierk Rainer).

Annotated Bibliography

We include a book in this annotated bibliography either because we found it helpful in writing *On the Trail of Blackbody Radiation* or because we recommend it for further study or both.

Callen, H. B. *Thermodynamics*. 1st ed. New York: Wiley, 1960. 376 pages. Studying this highly regarded postulational approach to classical thermodynamics, along with a historically based presentation of thermodynamics, is a good way to learn thermodynamics.

Carnot, Sadi. *Reflections on the Motive Power of Fire*. Gloucester, MA: Peter Smith, 1977. 192 pages. Translated by R. H. Thurston. In this, the first significant contribution to theoretical thermodynamics, Carnot constructs the second law of thermodynamics, without assuming the truth of the first law. The first part of Carnot's text is very readable.

Clausius, Rudolf. *The Mechanical Theory of Heat*. London: John van Voorst, 1867. 396 pages. A compendium of nine papers by Clausius published between 1850 and 1865 that established theoretical thermodynamics.

Darrigol, Olivier. *From c-Numbers to q-Numbers: The Classical Analogy in the History of Quantum Theory*. Berkeley: University of California Press, 1992. 381 pages. The first part of Darrigol's book (pp. 1–78) covers the same territory as the current text. However, Darrigol focuses on some issues, such as the debate over the reversibility or irreversibility of thermodynamic processes, that are beyond the scope of our book and ignores others, such as Boltzmann's derivation of the Stefan-Boltzmann law, that we discuss in detail.

Duncan, Anthony, and Michael Janssen. *Constructing Quantum Mechanics.* Oxford: Oxford University Press, 2019. 768 pages. A major and much-needed contribution to the subject. The authors briefly discuss Planck's derivation in the first few pages of the first volume of this planned two-volume work.

Griffiths, David J. *Introduction to Electrodynamics.* Englewood Cliffs, NJ: Prentice-Hall, 1981. 479 pages. A deservedly popular undergraduate textbook in which Griffiths derives the radiation reaction equation (section 7.10) from the non-Newtonian-third-law behavior of the electromagnetic interaction.

Heilbron, J. L. *The Dilemmas of an Upright Man.* Cambridge, MA: Harvard University Press, 1996. 272 pages. A biography of Planck that, while in many ways interesting, sheds little light on his derivation of the spectral energy density. Instead, Heilbron emphasizes Planck's contributions to the German scientific bureaucracy after 1900–1901. Planck's life was self-sacrificial and tragic. His first wife, Marie, and their four children all predeceased him. Marie died of tuberculosis at the age of 48, a son Karl was killed in World War I, his two daughters, Margaret and Emma, died in childbirth, and the Nazis executed his son Erwin for plotting to assassinate Hitler. His house, library, and papers were destroyed in an Allied bombing raid during World War II. Planck died in 1947.

Helrich, Carl. *Quantum Theory: Origins and Ideas: A Historical Primer for Physics Students.* Cham, Switzerland: Springer, 2021. 251 pages. An admirable, student-oriented introduction to the subject by an accomplished scholar and teacher.

Jammer, Max. *The Conceptual Development of Quantum Mechanics.* New York: McGraw-Hill, 1966. 436 pages. The first chapter of Jammer's book entitled "The Formation of Quantum Concepts" covers much the same ground as the current text. Jammer's scholarship is impeccable. For instance, his first chapter has 248 endnotes! However, his conceptual approach does not include derivations.

Kuhn, Thomas S. *Black-Body Theory and the Quantum Discontinuity 1894–1912*. Chicago: University of Chicago Press, 1987. 398 pages. This book is the single best secondary source on Planck's derivation of the spectral energy density of blackbody radiation. We consulted it frequently while writing our book. Kuhn's thesis could be characterized as, "Planck did not quantize anything."

Lemons, Don S. *Mere Thermodynamics*. Baltimore: Johns Hopkins University Press, 2009. 222 pages. This historically oriented introduction to classical thermodynamics emphasizes the subject's intellectual structure.

Planck, Max. *Eight Lectures on Theoretical Physics*. New York: Columbia University Press, 1915. 90 pages. Translated by A. P. Wills. Planck's fifth lecture (pp. 52–62) outlines his derivation of the "fundamental relation."

Planck, Max. *Planck's Original Papers in Quantum Physics*. London: Taylor and Francis, 1972. 60 pages. Edited and annotated by Hans Kangro. Translated by D. ter Haar and Stephen G. Brush. This dual-language text contains English translations of two of Planck's original papers: his October 1900 paper "On an Improvement of Wien's Equation for the Spectrum" and his January 1901 paper "On the Theory of the Energy Distribution Law of the Normal Spectrum."

Planck, Max. *A Scientific Autobiography*. New York: Philosophical Library, 1949. 196 pages. Translated by Frank Gaynor. A verbal recapitulation of Planck's scientific contributions including his work in thermodynamics and his derivation of the spectral energy density of blackbody radiation.

Planck, Max. *A Survey of Physical Theory*. Mineola, NY: Dover Publications, 1960. 117 pages. Translated by R. Jones and D. H. Williams. The lecture "The Origin and Development of the Quantum Theory" presents Planck's view of his subject circa 1920. His remarks are consistent with the outline of our text and with our verbal summary in "The Big Ideas." Planck summarizes the evidence for the "quantum of action" *h* (now known as Planck's constant) and admits that the existence of *h* does not constitute a theory of quantum physics and that such a theory is needed.

Planck, Max. *Theory of Heat Radiation*. 2nd ed. New York: Dover, Publications, 1991. 224 pages. Translated by Morton Masius. This text, first published in 1912, presents what Kuhn and other historians refer to as Planck's "second derivation" of the spectral energy density.

Planck, Max. *Treatise on Thermodynamics*. New York: Dover Publications, 1969. 297 pages. Translated by Alexander Ogg. First published in 1897. The publication date of this text is evidence that Planck was immersed in thermodynamic thinking while he derived the spectral energy density of blackbody radiation. It is also notable for Planck's precise statement of Nernst's theorem and its transformation into the third law of thermodynamics.

Richtmyer, F. K., and E. H. Kennard. *Introduction to Modern Physics*. New York: McGraw Hill, 1942. 723 pages. An example of an older, influential undergraduate physics text that presents the history of Planck's derivation of the spectral energy density of blackbody radiation.

Appendix A English Translation of "A Derivation of Stefan's Law,[1] Concerning the Temperature Dependence of Thermal Radiation, from the Electromagnetic Theory of Light" by Ludwig Boltzmann in Graz (1884)[2]

Maxwell concluded from his electromagnetic theory of light[3] that a beam of visible light or radiant heat normally incident on a unit area of surface must exert a force on it equal to the beam's radiant energy contained per unit volume of aether. Let us imagine, then, a completely empty volume surrounded by walls of absolute temperature t and impermeable to thermal radiation. If we denote by $\psi(t)$ the radiant energy now contained per unit volume of aether inside, we must bear in mind that not all radiant beams will meet the walls at normal incidence. It is simplest to take a point of view in analogy to that applied by Krönig[4] in his study of gases. We envision our volume as a cube whose sides are oriented along three mutually perpendicular axes. One achieves a result most closely corresponding to the true average by assuming that a third of the thermal radiation proceeds parallel to each of the axes. Each bounding surface experiences just a third of the radiation and the corresponding force on each unit area according to Maxwell's law is then:

$$f(t) = \frac{1}{3}\psi(t).$$

This result also follows from the following consideration. Maxwell's result is valid whenever the radiant beam encounters the surface normally and is completely absorbed. Were it to encounter the surface at near-normal incidence but then be reflected at the same angle, the resulting force would be doubled. Further, were it to meet the surface while forming an angle θ with the normal and then be reflected at that very angle with the same intensity, then, by analogy with an obliquely impinging gas molecule, only the normal component of the beam's full kinetic force will be brought to bear. This requires a combined factor of $\cos^2\theta$ since both the momentum of motion as well as the impinging quantity delivered both receive a $\cos\theta$ factor. If we once again denote the total radiant energy of all beams in a unit volume as $\psi(t)$, then the radiant energy of just those incoming beams forming an angle with the normal of the impinged surface between θ and $\theta + d\theta$ equals $\frac{1}{2}\psi(t)\sin\theta d\theta$ and the radiant energy of those outgoing beams at the same angle is of equal size, for a total of $\psi(t)\sin\theta d\theta$. Together these beams exert a total force equal to $\psi(t)\cos^2\theta\sin\theta d\theta$ on the unit area impacted. To include all participating beams we integrate this from zero to $\pi/2$ and arrive at $\psi(t)/3$ just as before.

In my essay on the result discovered by Bartoli concerning a relationship between radiant heat and the second law,[5] I showed that the second law further implies the following relationship between the two functions ψ and f:[6] $f = t\int\psi dt/t^2$. The differential of this result yields $tdf - fdt = \psi dt$ which then, when combined with the result from above, $f = \psi/3$, yields $td\psi/3 = 4\psi dt/3$. Upon integration this results in $\psi = ct^4$, a law which Stefan proposed empirically some time ago and which is found to be in good agreement with measurements. Therefore, Stefan's law concerning the dependence of radiant energy on temperature follows immediately from the electromagnetic theory of light and the second law. This is certainly a remarkable result even though no one will overlook the decidedly provisional character of the derivation in many respects.

Further, it follows easily by reversing the argument and now starting with both Stefan's law and the second law that, inside a volume surrounded by walls impermeable to thermal radiation and of uniform temperature, each unit planar surface must experience a force equal to a third of the contained radiant energy per unit volume. That is, a beam that falls at normal incidence on a unit planar surface and is then absorbed must exert a force on it equal to the energy per unit volume of the beam. In contrast, if the beam is reflected with undiminished intensity, then the force is twice as great and consequently equal to the conjoined energy contained per unit volume of the incident and reflected beams taken together. According to the emissions theory, as it seems to me, but in contrast to the view of Hirn as advanced by Bartoli, the force exerted would, in each of these cases, have to be twice as great as the values found here. Moreover, it seems to me that, by using hypotheses similar to those employed by Kirchhoff in his well-known treatise on the equality of the capacity for emission and absorption,[7] one could also prove that the law found here for completely black bodies applies not only to the total body of outgoing thermal radiation taken as a whole, but also applies to each subcategory of radiation individually. From this we would then have to conclude that, for completely black bodies, the emitted heat in any radiant subcategory is the same fraction of the total emitted heat at all temperatures, as suspected by Lecher[8] among others.

I mention, finally, a somewhat simpler derivation of the relationship between ψ and f as found from the Bartoli process. Let us imagine a hollow cylinder lying horizontally and thermally isolated by an absolutely black covering impermeable to heat. The cylinder is of unit cross-sectional area and its extreme left end, B, is closed and held at absolute temperature t_0. Imagine, further, a movable piston S, of the same description as the side walls, which is, at the outset, placed at the extreme left end of the cylinder and then is drawn to the right a distance a. The whole of the radiant heat energy now found between B and S, $a \cdot \psi(t_o)$, as well the heat expended in moving S (measured in work units), $a \cdot f(t_o)$, is delivered by B. We now thermally isolate the left end with a second piston T and proceed to draw piston S a distance x further adiabatically. The cylinder's walls and pistons are imagined to hold vanishingly small amounts of heat. During the second change of state we may write: $d[(a+x)\psi(t)] = -f(t)dx$ at each step until the left-hand volume finally attains a temperature t. The volume to the right of S is closed by a facing surface G at the right end of the cylinder which is held at that same final absolute temperature t throughout. All the heat extracted by the work performed on the right-hand side as well as the heat extracted by the diminished right-hand volume, $(a+x)[\psi(T)+f(t)]$, is taken up by the right-hand facing surface G. The process is reversible. Accordingly, we have:

$$(a+x)[\psi(t)+f(t)]/t = (a)[\psi(t_o)+f(t_o)]/t_o = c.$$

We now wish to view a and t_o as well as c as constants, and x and t as variables. If we add $d[(a+x)f(t)]$ to both sides of the previous equation and combine it with the differential of this last equation, there results:

$$\psi(t)dt + f(t)dt = tdf(t)$$

which agrees with the result from above.

Appendix B English Translation of "A New Relationship between Blackbody Radiation and the Second Law of Thermodynamics" by Willy Wien in Charlottenburg (1893)[1]

(submitted by Mr. H. von Helmholtz)

Boltzmann[2] proved, on the basis of a process devised by Bartoli, that the second law of thermodynamics implies the existence of a pressure which is exerted by any radiant beam on a surface it illuminates. Just such a pressure is also a consequence of the electromagnetic theory of light, and Boltzmann was able to show from this relationship that Stefan's radiation law for blackbodies follows.

These conclusions admit further completion if one views the radiant beam not merely as a whole but, rather, imagines decomposing it by wavelength.

The imagined processes which form the foundation of our considerations here must, as with Boltzmann and even earlier with Kirchhoff and Clausius, correspond so closely to reality that they may actually be implemented to an arbitrarily good approximation.

Among our preliminary assumptions we must first include the validity of the electromagnetic theory of light according to which a light beam exerts a pressure along its given direction equal to its energy. Next, we must admit the possibility of the existence of completely black and also completely reflective bodies that are so constituted as to reflect back any incident light in a totally diffuse manner such as we find well approximated by the complete reflection of white bodies. Furthermore, we view the second law of thermodynamics as valid so that, even by radiation processes proceeding from the internal heat reservoir of a solid body, no work can be gained without other attendant changes in work output, temperature, and thermodynamic state. Finally, we presume the validity of the Doppler principle in relation to light beams.

Blackbody radiation in an enclosed volume with reflective walls proceeds in all possible directions. Following Boltzmann, one arrives at the best average value corresponding to these conditions with the assumption that, in a cube, just one third of the total radiation proceeds parallel to each side wall. A reduction in volume results in a greater energy density (the energy per unit volume) both through containing the energy at hand in a smaller volume and through the added work performed against the radiation pressure. By reversing this process one recovers the added work completely if all volume changes have been performed at speeds vanishingly small compared to the speed of light, so that the density remains uniform throughout the volume.

A process may now be imagined whereby a specified energy density increase could be effected either by increasing the temperature or, alternatively, through reducing the volume.

The second law now indicates that with equal total energy densities for these two situations, it must also be equal for each wavelength individually. For the case of volume reduction, the wavelengths are altered in accordance with the Doppler principle, and this change may be calculated. The change effected through temperature increase may thereby be known.

§I Process Description

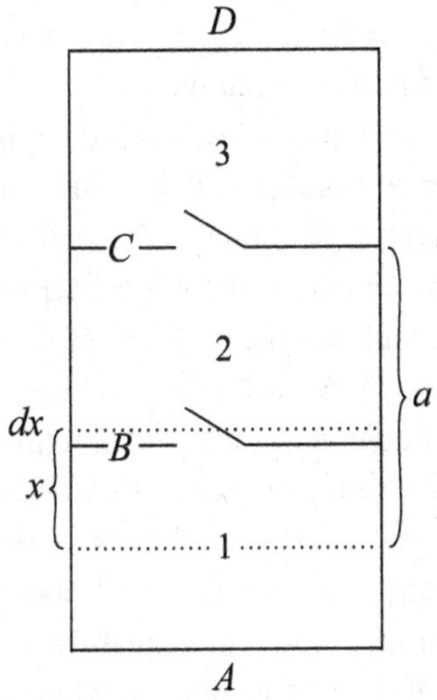

Let us imagine a cylinder of unit cross-sectional area in which we place two movable pistons denoted ***B*** and ***C***. Each of these pistons is to be equipped with a light-tight trapdoor. The pistons partition the cylinder into three subvolumes labeled 1, 2, 3 which are assumed to be completely empty. The end caps, ***A*** and ***D***, are blackbodies of different absolute temperatures ϑ_1 and $\vartheta_2 > \vartheta_1$. These end caps are imagined to be of such dimension that they may be considered heat reservoirs of arbitrarily large extent in comparison with the radiant energy content of the cylinder partitions. All the inner surfaces of the cylinder, the pistons, and their trapdoors shall reflect any incident radiation perfectly diffusively so that no preferential direction occurs. If one acknowledges that perfectly reflecting surfaces can be imagined, then certainly no objection will be raised against the possibility of producing such inner surfaces arbitrarily well and which we may now denote as completely white. Should the trapdoors of both pistons ***B*** and ***C*** be closed, then subvolume 2 entirely isolates the radiation within its walls which then remains unaltered.

At the outset of our process we open the trapdoor of ***B*** while keeping ***C*** sealed. End cap ***A*** irradiates subvolumes 1 and 2 and end cap ***D*** irradiates subvolume 3. The density of energy in 3 is greater than in 2 because ***D*** has a higher temperature than ***A***. We now close and seal the trapdoor of ***B*** and then proceed to displace piston ***B*** toward ***C*** with a velocity ***v*** which is vanishingly small compared to the speed of light ***c***. The energy in subvolume 2 will attain a greater density because of its diminished volume as well as because of the work performed against the pressure within. We continue this displacement until the densities in volumes 2 and 3 are equal. From the second law of thermodynamics we may now conclude that the energy distributions across the entire radiant spectrum must also be equal. Because, if this were not the case, then there would have to be at least one wavelength whose energy in subvolume 3 was greater than the energy found in 2. We could then position a thin filter across the trapdoor of ***C*** which allowed this wavelength to pass but reflected all others. More energy would then pass from 3 to 2 than in reverse, and the density of energy in 2 would become greater than in 3. We now close trapdoor ***C***, remove the filter, and allow the excess pressure in 2 to move piston ***C***, yielding work done until the energy densities are again in equality. Let the work thereby extracted be denoted by ***Q***. We now open the trapdoor of ***C*** once again and bring the piston back to its original position, which can be achieved without expenditure of work since the pressure is equal on both sides. Finally, upon closing the trapdoor of ***C*** once more, we may displace piston ***B*** back to its original position and recover all the work we expended when we moved it forward. When the trapdoor of ***B*** is finally reopened we will have completely recovered our original configuration and will have extracted work ***Q*** from heat without any compensating change of state. Since this would violate the second law, the spectral energy distribution in volumes 2 and 3 must have been the same. Accordingly, if we know the energy distribution in the radiant energy originating from ***A*** and the alteration it suffers as a consequence of the movement of piston ***B***, then we will also know the distribution in the radiation from the warmer body ***D***.

§2 Calculation of the Energy Distribution's Alteration Using the Doppler Principle

Let us, once again, denote the velocity of the piston by ***v*** and the velocity of light by **c**. According to the Doppler principle, radiation incident on piston ***B*** during its motion will have its wavelength diminished. The vibrational period T of any beam which is normally incident and suffers a single reflection will be altered according to the equation:

$$T' = \frac{c-2v}{c}T.$$

Since $T = \lambda/c$ and $T' = \lambda'/c$ where λ and λ' denote the corresponding wavelengths, then we also have:

$$\lambda' = \frac{c-2v}{c}\lambda.$$

In the event of skew incidence only the normal component comes into consideration. When the radiation is distributed uniformly in all directions in a cube over piston B, we may equate the sum total of the normally incident components to one third of the total energy. Those rays reflected from the moving piston will thereupon be diffusely scattered from the stationary walls they encounter, and directional uniformity will, in this manner, be immediately restored.

Let us denote by $\varphi(\lambda)$ the energy density originally present in subvolume 2 given as a function of wavelength, so that the energy for wavelengths lying between λ and $\lambda + d\lambda$ is $\varphi d\lambda$. After a single reflection from the moving piston the normally incident beams are shortened by an amount h and the new energy distribution may be denoted as $f_1(\lambda)$. If we imagine entering the quantity $\varphi(\lambda)$ as ordinate and λ as abscissa, then we will recover a point on the new curve $f_1(\lambda)$ if we retain for each λ two-thirds of the original ordinate $\varphi(\lambda)$ corresponding to the two-thirds portion of the energy remaining unaltered. The last third must be replaced by a third of the ordinate φ corresponding to $\lambda + h$, since a third of this energy has had its wavelength diminished by h. We have, consequently:

$$f_1(\lambda) = \frac{2}{3}\varphi(\lambda) + \frac{1}{3}\varphi(\lambda + h).$$

Whenever h is small we may write:

$$\varphi(\lambda + h) = \varphi(\lambda) + h\varphi'(\lambda)$$

and therefore

$$f_1(\lambda) = \varphi(\lambda + h/3)$$

after a single reflection from piston B. After an n-fold reflection we have, accordingly,

$$f_n(\lambda) = \varphi(\lambda + nh/3) = f(\lambda)$$

if nh remains small in comparison to λ.

From these it follows, again, that:

$$f(\lambda) = \varphi(\lambda) + \frac{nh}{3}\varphi'(\lambda).$$

The alteration in the energy density proceeds as if the normally incident rays, constituting a third of the total energy, had diminished their wavelength by the amount *nh*. We have now to conceive of the number *n* as the total count of the number of times normally incident rays are repelled back and forth from the moving piston in subvolume 2 while it advances a specific distance.

If $a - x$ specifies the distance from B to C, then while B advances by *dx* we have:

$$n = \frac{dx}{2(a-x)} \cdot \frac{c}{v}$$

assuming that *n* is large in comparison to unity. We must have, then, that c/v remains large in comparison to $2(a-x)/dx$ while *dx* remains small compared to $2(a-x)$.

If, after a single reflection, we concluded that

$$\lambda' = \frac{c-2v}{c}\lambda$$

then an *n*-fold repetition yields

$$\lambda_n = \left(\frac{c-2v}{c}\right)^n \lambda$$

$$= \left(\frac{c-2v}{c}\right)^{\frac{dx}{2(z-x)}\cdot\frac{c}{v}} \lambda.$$

This allows a rearrangement as

$$\lambda_n = \left[\left(1-\frac{2v}{c}\right)^c\right]^{\frac{dx}{2(a-x)}\cdot\frac{1}{v}} \lambda.$$

From this we take the limit $c = \infty$ and achieve

$$\lambda_n = \left[e^{-2v}\right]^{\frac{dx}{2(a-x)}\cdot\frac{1}{v}} \lambda = e^{\frac{-dx}{a-x}}\lambda.$$

One observes that on the piston's return stroke we would recover

$$\lambda = \left(\frac{c+2v}{c}\right)^n \lambda_n = e^{\frac{dx}{a-x}} \lambda_n$$

and that λ_n would return to its original value. The process is therefore reversible. We now set

$$\lambda_n = \lambda + nh$$

where nh is infinitesimal and of the order of dx. Neglecting small terms of second order we may now write

$$nh = -\frac{dx}{a-x}\lambda.$$

And finally, combining this with our results from above, we achieve

$$f(\lambda) = \varphi(\lambda + nh/3) = \varphi(\lambda + d\lambda)$$

where we may identify:

$$d\lambda = -\frac{dx}{3(a-x)}\lambda.$$

Every value of λ is diminished by this amount if x increases by dx.

By integration we obtain:

$$\ln\lambda = \frac{1}{3}\ln(a-x) + \ln C$$

$$\ln C = \ln\lambda_o - \frac{1}{3}\ln a$$

where λ_o is the value for $x = 0$. Therefore, we have:

$$\lambda = \sqrt[3]{\frac{a-x}{a}} \cdot \lambda_o.$$

Now let us denote by E the quantity of energy in subvolume 2. The density of energy is then:

$$\psi = \frac{E}{a-x}.$$

Upon x growing by dx, the energy density must increase by dint of the diminished volume of containment as well as by the work performed on it.

$$\begin{aligned}\frac{d\psi}{dx}dx &= \left\{\frac{dE}{dx}\frac{1}{a-x} + \frac{E}{(a-x)^2}\right\}dx \\ &= \left(\frac{dE}{dx} + \psi\right)\frac{dx}{a-x}.\end{aligned}$$

Recalling that the pressure on the piston is assumed to be given by

$$= \frac{1}{3}\psi$$

then the work performed must be

$$\frac{dE}{dx}dx = \frac{1}{3}\psi dx.$$

This yields

$$\begin{aligned}d\psi &= \frac{4}{3}\frac{\psi}{a-x}dx \\ \ln\psi &= \ln\left[(a-x)^{4/3}\right] + \ln C_1 \\ \ln C_1 &= \ln\psi_o + \ln\left(a^{4/3}\right)\end{aligned}$$

where ψ_0 is the value for $x = 0$.

We conclude, therefore, that:

$$\psi = \sqrt[3]{\left(\frac{a}{a-x}\right)^4} \cdot \psi_o.$$

And, finally, combining this with our earlier equality, we have for equal x values that:

$$\frac{\psi}{\psi_0} = \frac{\lambda_0^4}{\lambda^4}.$$

Now, by the argument of §1, the distribution of the energy represented by ψ must be identical to that originating from a higher temperature ϑ. If ϑ_0 is the value of ϑ corresponding to ψ_0, then by the Stefan-Boltzmann result we must have:

$$\frac{\psi}{\psi_0} = \frac{\vartheta^4}{\vartheta_0^4}$$

and it follows that

$$\vartheta\lambda = \vartheta_0\lambda_0.$$

Under a temperature alteration, every wavelength in the normal emissions spectrum of a blackbody suffers a displacement such that the product of the temperature and the wavelength remains constant.

If the energy distribution is specified as a function of wavelength for any one temperature ϑ_0, then we may now derive it for all other temperatures ϑ. Let us, once again, imagine entering the quantity λ as abscissa and $\varphi(\lambda)$ as ordinate. The area contained between the presented curve and the abscissa axis represents the total energy ψ. We have next so to alter every λ that $\vartheta\lambda$ remains constant. If we were to cut out a narrow band of width $d\lambda_0$ at the original position λ_0 with areal content $\varphi_0 d\lambda_0$, then this would have displaced to the position λ and from the width $d\lambda_0$ would have arisen the width $d\lambda_1 = \frac{\vartheta_0}{\vartheta} d\lambda_0$. Since the quantity of energy $\varphi_0 d\lambda_0$ must remain constant, it follows that

$$\varphi d\lambda = \varphi_0 d\lambda_0, \varphi = \varphi_0 \frac{d\lambda_0}{d\lambda} = \varphi_0 \frac{\vartheta}{\vartheta_0}.$$

In addition, every φ is altered in accord with Stefan's law by the factor $\vartheta^4/\vartheta_o^4$, so that the new ordinate must be

$$\varphi = \varphi_o \frac{\vartheta^5}{\vartheta_o^5}.$$

In this manner, each point on the new distribution curve may be specified.

This result is in complete agreement with the conclusions of H. F. Weber[3] concerning the displacement of the energy maximum as derived from the laws of radiation.

Appendix C An Electromagnetic Adiabatic Invariant

We use a special system to identify the *electromagnetic adiabatic invariant*. Accordingly, consider a slowly shrinking sphere of radius r from whose inside surface isotropic, monochromatic radiation reflects. Each beam or pulse of radiation propagates within a plane cross-section of the sphere that contains the sphere center[1]—as shown in Figure C.1. This monochromatic radiation is Doppler-shifted by an amount $\Delta\lambda_1$ given by

$$\frac{\Delta\lambda_1}{\lambda} = 2\cos\phi \frac{dr/dt}{c} \tag{C.1}$$

as it reflects from the inner surface of the sphere, where $\Delta\lambda_1$ indicates the Doppler shift due to one reflection. Also, c is the speed of light, the factor $\cos\phi$ accounts for the angle ϕ to the normal of the surface at which the radiation reflects, and $|dr/dt| << c$ so that the Doppler shift is nonrelativistic.

After reflecting once from the surface, the radiation travels a distance x where

$$\begin{aligned} x^2 &= 2r^2 - 2r^2\cos(\pi - 2\phi) \\ &= 2r^2(1 + \cos 2\phi) \\ &= 4r^2\cos^2\phi \end{aligned} \tag{C.2}$$

at which point the radiation reflects again. The distance x between reflections is $2r\cos\phi$, the period between reflections

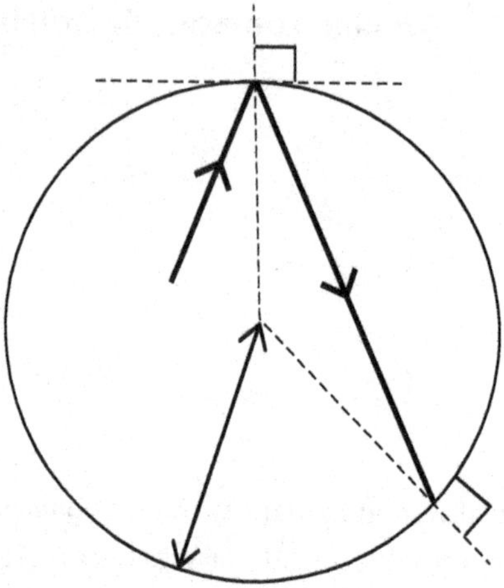

Figure C.1.
Electromagnetic radiation of wavelength λ reflecting from the interior surface of a sphere whose radius r is shrinking at speed dr/dt.

is x/c, that is, $2r\cos\varphi/c$, and the frequency of reflection is $c/2r\cos\varphi$. Therefore, the rate at which the wavelength Doppler-shifts is given by

$$\begin{aligned}\frac{d\lambda}{dt} &= \left(\frac{c}{2r\cos\phi}\right)\Delta\lambda_1 \\ &= \left(\frac{c}{2r\cos\phi}\right)\left(\frac{2\lambda\cos\phi}{c}\frac{dr}{dt}\right) \\ &= \frac{\lambda}{r}\frac{dr}{dt}.\end{aligned} \tag{C.3}$$

Since (C.3) is observed for all values of ϕ and λ in all planes intersecting the center of the sphere, it is observed for all radiation within the sphere. Equation (C.3) is solved by

$$r = p\cdot\lambda \tag{C.4}$$

where p is a dimensionless constant independent of all thermodynamic variables. Given $V = 4\pi r^3/3$, we find that

$$V^{1/3} = \left(\frac{4\pi}{3}\right)^{1/3} p\lambda \\ = q \cdot \lambda \tag{C.5}$$

where q is a dimensionless constant that describes all such systems with volume V and monochromatic, blackbody radiation of wavelength λ. Thus,

$$\frac{V^{1/3}}{\lambda} = ad.inv. \tag{C.6}$$

describes the electromagnetic adiabatic invariant $V^{1/3}\lambda^{-1}$.

Appendix D An Ideal Gas "Displacement Law"

Parallels between the ideal gas and blackbody radiation extend more deeply than has yet been mentioned. Suppose we are now in the position with respect to the ideal gas that Wien was with respect to blackbody radiation in 1893. For instance, suppose we know the two independent equations of state of the ideal gas but not the quantity corresponding to the spectral energy density of blackbody radiation, that is, not the distribution of particle speeds known as *Maxwell's distribution.* We find that we can take the same steps in the same order with reference to the ideal gas that we have taken in chapter 5 in deriving the Wien displacement law. Our result will be a formal restriction on the distribution of ideal gas particle speeds that we call the *ideal gas displacement law.*

First recall that the two independent equations of state of the ideal gas are $PV = nRT$ and $U = C_V T$. Nothing is lost and a few expressions are simplified if we immediately specialize to a monatomic ideal gas. In this case, we know that $C_V = 3nR/2$ where n is the number of moles of gas. We then define a *cumulative energy density* $U_0^v(T)$ with units of energy per volume where

$$U_0^v = \left(\frac{3nRT}{2V}\right)\int_0^v f_v(T,v)dv \tag{D.1}$$

and $f_v(T,v)dv$ is the unknown fraction of particle speeds between v and $v+dv$ of an ideal gas with temperature T. The $v\to\infty$ limit of definition (D.1),

$$\int_0^\infty f_v(T,v)dv = 1, \tag{D.2}$$

places a restriction on the integral of the speed distribution $f_v(T,v)$. Furthermore, the relation

$$\frac{\partial U_0^v}{\partial v} = \left(\frac{3nRT}{2V}\right) f_v(T,v) \tag{D.3}$$

allows us to look under the integral sign of definition (D.1).

The fundamental constraint for a system of ideal gas particles with speeds between 0 and v is therefore

$$d\left(U_0^v V\right) = TdS_0^v - P_0^v dV \tag{D.4}$$

where S_0^v and P_0^v are the system's entropy and isotropic pressure and T and V are its temperature and volume. Since the pressure of the ideal monatomic gas is, in terms of the independent variables U and V, equal to $P = 2U/3V$, the pressure of the system of ideal gas particles with speeds between 0 and a cutoff speed v is given by

$$P_0^v = \frac{2}{3} U_0^v. \tag{D.5}$$

Recall that U_0^v is an energy density while U is an energy.

The adiabatic invariant of this system follows from the fundamental constraint (D.4), the equation of state $P_0^v = 2U_0^v/3$, and $dS_0^v = 0$. Then the fundamental constraint (D.4) reduces to

$$VdU_0^v + U_0^v dV = -\frac{2}{3} U_0^v dV \tag{D.6}$$

from which follows

$$VdU_0^v + \frac{5U_0^v}{3}dV = 0. \tag{D.7}$$

The adiabatic invariant

$$U_0^v V^{5/3} = ad.inv. \tag{D.8}$$

solves (D.7). Recall that *ad.inv.* indicates any combination of variables that remains constant during an isentropic or, equivalently, a reversible, adiabatic process with no dissipation or heat transfer. For this reason, we know that the functionality of the entropy is reduced from two variables $S_0^v(U_0^v V, V)$ to one variable $S_0^v(U_0^v V^{5/3})$.

The fundamental constraint (D.4) also implies that

$$\left[\frac{\partial S_0^v}{\partial(U_0^v V)}\right]_V = \frac{1}{T}. \tag{D.9}$$

Therefore, given the identity $[\partial S_0^v(U_0^v V^{5/3})/\partial(U_0^v V)]_V = S_0^{v\prime}(U_0^v V^{5/3})V^{2/3}$ where $S_0^{v\prime}(U_0^v V^{5/3})$ indicates a derivative with respect to argument, it follows from (D.9) that

$$S_0^{v\prime}(U_0^v V^{5/3}) = \frac{1}{V^{2/3}T}. \tag{D.10}$$

Since a function of an adiabatic invariant is an adiabatic invariant,

$$V^{2/3}T = ad.inv. \tag{D.11}$$

is another form of the adiabatic invariant of this system. Thus, $U_0^v V^{5/3}$ and $V^{2/3}T$ are two forms of the thermodynamic adiabatic invariant of the system of ideal gas particles with speeds between 0 and v.

In addition, the speeds v of the particles in the system observe a *mechanical adiabatic invariant*. Imagine, for instance, an ideal gas particle with mass m and speed v that smoothly and elastically reflects from the inside of a spherical container of radius r that, in turn, slowly and self-similarly shrinks in such a way that its radius slowly diminishes at a constant rate $\dot{r}[<0]$ so that $|\dot{r}| << v$. Figure C.1, which illustrated the specular reflection of light, also illustrates the geometry of successive particle reflections from the inside of a smooth spherical container. At each reflection the particle gains a speed

$$\Delta v = -2\dot{r}\cos\varphi \tag{D.12}$$

where ϕ is the particle's angle of incidence and reflection with respect to the surface normal. Therefore, the average force F exerted on the particle by the spherical container is given by

$$\begin{aligned} F &= \frac{m\Delta v}{\Delta t} \\ &= -\frac{m2\dot{r}\cos\phi}{\Delta t} \end{aligned} \tag{D.13}$$

where Δt is the interval between collisions when the angle of reflection is ϕ. Accordingly,

$$\Delta t = \frac{2r\cos\varphi}{v} \tag{D.14}$$

so that this force is given by

$$\begin{aligned} F &= -m\frac{2\dot{r}\cos\phi}{2r\cos\phi/v} \\ &= -\frac{m\dot{r}v}{r}. \end{aligned} \tag{D.15}$$

Newton's second law, according to which this force is equal to mdv/dt, and (D.15) together imply that

$$\frac{dv}{dt} = -\frac{\dot{r}}{r}v \tag{D.16}$$

whose solution is

$$vr = ad.inv.. \tag{D.17}$$

The quantity vr is an adiabatic invariant since the process of reflection is reversible and no heat is transferred into or out of the system. The container volume V is related to its radius r by $V = 4\pi r^3/3$. Hence, (D.17) is equivalent to

$$V^{1/3}v = ad.inv.. \tag{D.18}$$

This adiabatic invariant applies to all the particles of ideal gas with speeds between 0 and v. To summarize, the quantities $U_0^v V^{5/3}$, $V^{2/3}T$, and $V^{1/3}v$ are adiabatic invariants of the system of ideal gas particles with speeds between 0 and v.

Since the quotient of two adiabatic invariants is also an adiabatic invariant, it follows that

$$\frac{U_0^v V^{5/3}}{V^{2/3}T} = \frac{U_0^v V}{T} = ad.inv. \tag{D.19}$$

and

$$\frac{\left(vV^{1/3}\right)^2}{V^{2/3}T} = \frac{v^2}{T} = ad.inv.. \tag{D.20}$$

Finally, since every single form of the adiabatic invariant is a function of every other form, we have

$$\frac{U_0^v V}{T} = g\left(\frac{v^2}{T}\right) \tag{D.21}$$

where $g(v^2/T)$ is an undetermined function.

The ideal gas "displacement law" is but a short step beyond (D.21). Taking the derivative with respect to v of both sides of (D.21), given (D.3), that is, given $(\partial U_0^v/\partial v)_T = f_v(T,v)$, produces

$$f_v(T,v) = \left(\frac{4v}{3nRT}\right)h\left(\frac{v^2}{T}\right) \tag{D.22}$$

where the undetermined function $h(v^2/T) = g'(v^2/T)$. Equation (D.22) is the ideal gas "displacement law."

This derivation of the ideal gas displacement law (D.22) is merely an exercise whose purpose is to further illustrate the method we use to derive Wien's displacement law in sections 5.2–5.5. After all, we already know that the fraction $f_v(T,v)dv$ of the speeds of ideal gas particles within an interval from v to $v + dv$ is given by

$$f_v(T,v)dv = \left(\frac{m}{2\pi kT}\right)^{3/2} e^{-mv^2/2kT}\, 4\pi v^2 dv. \tag{D.23}$$

This means that the function $h(v^2/T)$ must be

$$h\left(\frac{v^2}{T}\right) = \alpha e^{-\beta v^2/T} \sqrt{\frac{v^2}{T}} \tag{D.24}$$

where a and β are independent of the thermodynamic variables. Comparing (D.22), (D.23), and (D.24), we find that

$$\alpha = 4\pi \left(\frac{m}{2\pi}\right)^{3/2} \tag{D.25}$$

and

$$\beta = \frac{m}{2k}. \tag{D.26}$$

If one fit (D.22) and (D.24) to data on the speed distribution of an ideal monatomic gas, (D.25) and (D.26) would allow one to determine the value of the mass m of a gas particle and the value k of Boltzmann's constant—in a manner structurally similar to how Planck determined the value of Planck's constant h and Boltzmann's constant k.

Notes

Preface

1. Max Jammer, *The Conceptual Development of Quantum Mechanics* (New York: McGraw-Hill, 1966), 1.

2. See, for instance, the chapter "The Origin of the Quantum Theory" in F. K. Richtmyer and E. H. Kennard, *Introduction to Modern Physics* (New York: McGraw Hill, 1942).

3. Thomas S. Kuhn, *Black-Body Theory and the Quantum Discontinuity, 1894–1912* (Chicago: University of Chicago Press, 1987).

Chapter 1

1. This quotation and others concerning Pictet's experiment and its interpretations are from James Evans and Brian V. Popp's article "Pictet's Experiment: The Apparent Radiation and Reflection of Cold," *American Journal of Physics* 53 (1985): 737–753.

2. Pierre Prevost, "The Laws of Radiation and Absorption," in *Treatise on Radiant Heat, Scientific Memoirs XV*, translated and edited by D. B. Brace (New York: American Book Company, 1921), 19.

Chapter 2

1. Max Planck, *Treatise on Thermodynamics*, 3rd revised English edition, translated by A. Oog (New York: Dover Publications, 1969).

2. Paul Arthur Schilpp, ed. and trans., *Albert Einstein: Philosopher-Scientist* (La Salle, IL: Open Court, 1970), 33.

3. Rudolf Clausius, *The Mechanical Theory of Heat* (London: John van Voorst, 1867).

4. According to Arnold Sommerfeld it was Ralph H. Fowler who invented the term *zeroth law of thermodynamics*. See Arnold Sommerfeld, *Thermodynamics and Statistical Mechanics*, vol. 1, *Lectures on Theoretical Physics* (Cambridge, MA: Academic Press, 1951), 1. Also see Planck, *Treatise on Thermodynamics*, article 2.

5. Sadi Carnot, *Reflections on the Motive Power of Fire*, translated by R. H. Thurston (Gloucester, MA: Peter Smith, 1977).

6. William Thomson, "On an Absolute Temperature Scale, Founded on Carnot's Theory of the Motive Power of Heat, and Calculated from Regnault's Observations," *Philosophical Magazine* (October 1848). Thomson's paper also appears in W. F. Magie, *A Source Book in Physics* (Cambridge, MA: Harvard University Press, 1963), 237–242.

7. For a derivation of Clausius's theorem see Don S. Lemons, *Mere Thermodynamics* (Baltimore: Johns Hopkins University Press, 2009), 50–54.

Chapter 3

1. C. J. Adkins, *Equilibrium Thermodynamics*, 2nd ed. (Maidenhead, England: McGraw Hill, 1975), 151.

2. Quoted by J. L. Heilbron in *The Dilemmas of an Upright Man* (Cambridge, MA: Harvard University Press, 1996), 6.

Chapter 4

1. James Clerk Maxwell, *A Treatise on Electricity and Magnetism*, vol. I (New York: Dover Publications, 1954), 440–441.

2. Ludwig Boltzmann, "Ableitung des Stefan'schen Gesetzes, betreffend die Abhängigkeit der Wärmestrahlung von der Temperatur aus

der electromagnetischen Licttheorie," *Annalen der Physik und Chemie* 258, no. 6 (1884): 291–294. See Appendix A for an English translation by Louis Buchholtz.

3. Max Planck, *Theory of Heat Radiation*, translated by Morton Masius (Mineola, NY: Dover Publications, 1991), 62–63.

4. See D. S. Lemons, *Mere Thermodynamics* (Baltimore: Johns Hopkins University Press, 2009), 142ff., or any thermodynamics text that sketches the history and meaning of the Van der Waals equation of state.

5. Such as the one described in Planck, *Theory of Heat Radiation*, 62–63.

Chapter 5

1. Wien's actual contribution did not quite produce this form. See Appendix B for Louis Buchholtz's translation of Wien's 1893 paper, "Eine neue Beziehung der Strahlung schwarzer Körper zum zweiten Hauptsatz der Warmetheorie," *Sitzungsberichte der Königlich Preußichen Akademie der Wissenschaften zu Berlin* (1893): 55–62.

2. So named by Otto Lummer and Ernst Prigsheim in 1899. See Hans Kangro, *Early History of Planck's Radiation Law*, translated by R. E. W. Maddison (New York: Taylor & Francis, 1976), 45–46.

3. Max Planck, *The Theory of Heat Radiation*, translated by Morton Masius (Mineola, NY: Dover Publications, 1991), 69–86.

4. Edgar Buckingham, "On the Deduction of Wien's Displacement Law," *Bulletin of the Bureau of Standards* 8 (1912): 545–557, Scientific Paper 180 (S180).

5. Max Born, *Atomic Physics*, translated by John Dougall (London: Blackie and Sons, 1935), 239–240 and 408–412.

6. A. J. Adkins, *Equilibrium Thermodynamics* (Maidenhead, England: McGraw-Hill, 1975), 154–159.

7. See the third paragraph of Appendix A.

8. Planck, *Theory of Heat Radiation*, 89.

9. Albert Einstein, "Concerning an Heuristic Point of View toward the Emission and Transformation of Light," translated by A. B. Arons and A. B. Pippard, *American Journal of Physics* 33 (May 1965): 367–374. Originally published in *Annalen der Physik* 17 (1905): 132.

10. See, for instance, Adkins, *Equilibrium Thermodynamics*, 156–157. See also Appendix C, "An Electromagnetic Adiabatic Invariant."

11. Paul Ehrenfest, "Adiabatishe Invarianten and Quantentheorie," *Annalen der Physik* 51 (1916): 327–352. For an English translation see "Adiabatic Invariants and the Theory of Quanta," in *Sources of Quantum Mechanics*, edited by B. L. Van der Waerden (New York: Dover Publications, 1967), 79–93.

12. Thomas Kuhn, *Black-Body Theory and the Quantum Discontinuity, 1894–1912* (Chicago: University of Chicago Press, 1987), 86 and footnotes 24 and 30 on 277–278.

13. See *Planck's Original Papers in Quantum Physics*, edited and annotated by Hans Kangro, translated by D. ter Haar and Stephen G. Brush (London: Taylor and Francis, 1972), 52, annotation 19.

Chapter 6

1. Max Planck, *Eight Lectures on Theoretical Physics*, translated by A. P. Wills (New York: Columbia University Press, 1915), 59.

2. Paul Dirac, *The Principles of Quantum Mechanics*, 4th ed. (London: Oxford University Press, 1967), 58–61.

Chapter 7

1. Max Planck, "The Origin and Development of the Quantum Theory," in *A Survey of Physical Theory* (New York: Dover Publications, 1960), 103.

2. Thomas Kuhn, *Black-Body Theory and the Quantum Discontinuity, 1894–1912* (Chicago: University of Chicago Press, 1987), 83.

3. Thomas Kuhn placed the approach Planck took in his 1909 Columbia lectures squarely within the category of Planck's "first derivation." See Max Planck, *Eight Lectures on Theoretical Physics*, translated by A. P. Wills (New York: Columbia University Press, 1915), and Kuhn, *Black-Body Theory and the Quantum Discontinuity*, 306, note 45. Planck's derivation of the fundamental relation is also summarized on pp. 368–369 of Abraham Pais's excellent biography of Einstein, *Subtle Is the Lord: The Science and the Life of Albert Einstein* (Oxford: Oxford University Press, 1982).

4. Equation (7.10) necessarily follows from (7.9) only when the integrand on the right-hand side of (7.9) is known to be positive definite—a condition that here cannot be *a priori* guaranteed. Neither Planck in his *Eight Lectures on Theoretical Physics*, 59, nor J. D. Jackson when reviewing this process in *Classical Electrodynamics* (New York: Wiley, 1962), 581–582, take note of this problem. Jackson refers to the equivalent of (7.10) as the *Abraham-Lorentz equation of motion*, which he says "can be considered as an equation which includes in some approximate and time-average way the reactive effects of the emission of radiation." L. D. Landau and E. M. Lifshitz in *The Classical Theory of Fields*, 3rd ed. (Oxford: Pergamon Press, 1971), 203–206, also discuss the validity of the radiation reaction equation (7.10). See also D. J. Griffiths, *Introduction to Electrodynamics* (Englewood Cliffs, NJ: Prentice Hall, 1981), 380–383.

Chapter 8

1. In, for instance, Max Planck, *Vorlesungen über die Theorie der Wärmestrahlung* (Leipzig: J. A. Barth, 1906). Only Planck's second edition of this work has been translated into English; see Planck, *Theory of Heat Radiation*, translated by Morton Masius (1914; Mineola, NY: Dover Publications, 2011).

2. M. Planck, *Annalen der Physik* 1, no. 306 (1900): 730. For an English translation see Max Planck, "On an Improvement of Wien's Equation for the Spectrum," in *Planck's Original Papers in Quantum Physics*, edited and annotated by Hans Kangro, translated by D. ter Haar and Stephen G. Brush (London: Taylor and Francis, 1972), 35.

3. Max Planck's Nobel Prize address in *The Origin and Development of Quantum Theory*, translated by H. T. Clarke and L. Silberstein (Oxford: Oxford University Press, 1922), 6.

4. Planck, "On an Improvement of Wien's Equation for the Spectrum," 35.

5. Planck, "On an Improvement of Wien's Equation for the Spectrum," 37.

6. See, for instance, chapter 1 of William H. Cropper, *The Quantum Physicists and an Introduction to Their Physics* (London: Oxford University Press, 1970).

7. H. B. Callen, *Thermodynamics* (New York: Wiley, 1960), 9.

8. Planck's summary, on October 19, 1900, of the argument he made earlier in the year is so sketchy as to be almost useless. See Planck, "On an Improvement of Wien's Equation for the Spectrum," 35–36 and 48. Our reproduction of the essential points of Planck's argument follows Thomas Kuhn's somewhat more complete presentation in *Black-Body Theory and the Quantum Discontinuity, 1894–1912* (Chicago: University of Chicago Press, 1987), 95–96.

9. Max Planck, *Scientific Autobiography*, translated by Frank Gaynor (New York: Philosophical Library, 1949), 38–39. See also Max Planck, *A Survey of Physical Theory*, translated by R. Jones and D. H. Williams (New York: Dover Publications, 1960), 105; Kuhn, *Black-Body Theory and the Quantum Discontinuity*, 84–91.

10. Planck, "On an Improvement of Wien's Equation for the Spectrum," 36.

11. Kuhn, *Black-Body Theory and the Quantum Discontinuity*, 147 and 281, note 12.

12. Planck, *Scientific Autobiography*, 41.

Chapter 9

1. Ludwig Boltzman, "On the Relationship between the Second Fundamental Theorem of the Mechanical Theory of Heat and Probability

Calculations Regarding the Conditions for Thermal Equilibrium," translated by Kim Sharp and Franz Matschinsky, *Entropy* 17 (2015): 1971–2009, doi:10.3390/e17041971. The paper was originally published in *Kaiserlichen Akademie der Wissenschaften. Mathematisch-Naturwissen Classe. Abt. II* 76 (1877): 373–435, and was reprinted in Boltzmann, *Wissenschaftliche Abhandlungen*, vol. II, reprint 42 (Leipzig: Barth, 1909), 164–223.

2. Boltzmann illustrates the idea of equal probability with "the game of Lotto where every single quintet is as improbable as the quintet 12345."

3. For Boltzmann "log" meant what we refer to as "natural log" and sometimes symbolize with "ln."

4. See Boltzmann, "On the Relationship between the Second Fundamental Theorem of the Mechanical Theory of Heat and Probability Calculations Regarding the Conditions for Thermal Equilibrium."

5. Abraham Pais, *Subtle Is the Lord: The Science and the Life of Albert Einstein* (Oxford: Oxford University Press, 1982), 65–68.

6. Thomas S. Kuhn, *Black-Body Theory and the Quantum Discontinuity, 1894–1912* (Chicago: University of Chicago Press, 1987), 47–48. Planck was also conversant with Boltzmann's monograph *Lectures on Gas Theory* in which Boltzmann exploited techniques first introduced in 1877. See Ludwig Boltzmann, *Lectures on Gas Theory*, translated by Stephen G. Brush (Mineola, NY: Dover Publications, 2011).

7. Pais, *Subtle Is the Lord*, 60.

8. Kuhn, *Black-Body Theory and the Quantum Discontinuity*, 71.

9. Max Planck, *The Theory of Heat Radiation*, translated by Morton Masius (Mineola, NY: Dover Publications, 1991). See especially pp. 118–119.

10. Boltzmann, at first, specified a maximum index i above which $\omega_i = 0$ in order that he might address a particular finite problem, before allowing the index i to proceed to infinity. Of course, when

the index is allowed to proceed to infinity ω_i becomes very close to zero for all values of i such that $i\varepsilon > L$.

11. Kuhn, *Black-Body Theory and the Quantum Discontinuity*, 71.

Chapter 10

1. Max Planck, "On an Improvement of Wien's Equation for the Spectrum," as read to the German Physical Society on October 19, 1900, in *Planck's Original Papers in Quantum Physics*, edited by Hans Kangro, translated by D. ter Haar and Stephen G. Brush (London: Taylor & Francis, 1973), 35–37.

2. Thomas S. Kuhn, *Black-Body Theory and the Quantum Discontinuity, 1894–1912* (Chicago: University of Chicago Press, 1987), 97.

3. Max Planck, "On the Theory of the Energy Distribution Law of the Normal Spectrum," as read to the German Physical Society on December 14, 1900, in *Planck's Original Papers in Quantum Physics*, 38–45.

4. Max Planck, "On the Law of the Energy Distribution in the Normal Spectrum," translated by Yu. V. Kuyanov, http://theochem.kuchem.kyoto-u.ac.jp/Ando/planck1901.pdf (accessed December 15, 2018); originally published in German in *Annalen der Physik* 4 (January 7, 1901): 553ff.

5. Max Planck, Nobel Address, 1922, https://www.nobelprize.org/prizes/physics/1918/planck/lecture/ (accessed February 3, 2020).

6. Abraham Pais, *Subtle Is the Lord: The Science and the Life of Albert Einstein* (Oxford: Oxford University Press, 1982), 60.

7. Kuhn, *Black-Body Theory and the Quantum Discontinuity*, 98; a pointed summary of Planck's argument appears in an afterword on pp. 349–370. Martin J. Klein may have been the first to make this point. See his discussion in "Max Planck and the Beginnings of Quantum Theory," *Archive for the History of Exact Science* 1 (1961): 459–479.

8. A major thesis of Thomas Kuhn's magisterial book *Black-Body Theory and the Quantum Discontinuity* is that before the period

1908–1909 there is little evidence that Planck realized he had quantized anything. See the afterword of his book, especially pp. 350–351.

9. Planck, "On an Improvement of Wien's Equation for the Spectrum," 35–37.

10. Max Planck, *Vorlesungen über die Theorie der Wärmestrahlung* (Leipzig: J. A. Barth, 1906).

11. Kuhn makes sense of Planck's procedure in the *Annalen* paper by interpreting it as a short-cut way of considering all resonators including those of various frequencies. See Kuhn, *Black-Body Theory and the Quantum Discontinuity,* 102–110 and his summary in the afterword.

12. In adopting the phrase "elements of size ε" we deliberately mimic Planck's language in the January 7, 1901 *Annalen* paper.

13. Planck, "On an Improvement of Wien's Equation for the Spectrum," 36.

14. Thomas S. Kuhn makes this point. See Kuhn, *Black-Body Theory and the Quantum Discontinuity,* 97–101. Also see Klein, "Max Planck and the Beginning of Quantum Theory," 474.

15. Why Planck's combinatoric expression for the multiplicity (10.8) differs from Boltzmann's combinatoric formula (9.1) is a question discussed by historians. See, for instance, Clayton Gearhart, "Planck, the Quantum, and the Historians," *Physics in Perspective* 4 (2002): 170–215, especially 203–206. Boltzmann used variational methods to find the multiplicity of the stationary distribution or set of numbers $\{\omega_1, \omega_2, \omega_3, \ldots\}$, where ω_i is the number of resonators with energy $i\varepsilon$, subject to constraints on total energy and number of resonators as worked out in section 9.3. Then he identified this stationary distribution with equilibrium. As we have shown in section 9.4, this procedure leads to an entropy that is identical to that sought and found by Planck via the method described here. Planck, instead, used $S_N = k \ln \Omega$ where the multiplicity Ω is the permutation number of *all* the distributions satisfying system constraints. The two approaches produce the same result because the permutation number of the

equilibrium distribution is effectively the same as the permutation number of all distributions.

16. Planck, "On the Theory of the Energy Distribution Law of the Normal Spectrum," 42–43.

17. K. A. Tomilin, "Natural Systems of Units: To the Centenary Anniversary of the Planck System," *Proceedings of the XXII Workshop on High Energy Physics and Field Theory* (1999), 287–296; Max Planck, *The Theory of Heat Radiation*, translated by Morton Masius (Mineola, NY: Dover Publications, 1959), 173–175.

18. Tomilin, "Natural Systems of Units." Also see Don S. Lemons, *A Student's Guide to Dimensional Analysis* (Cambridge: Cambridge University Press, 2017), 85–95.

19. Kuhn, *Black-Body Theory and the Quantum Discontinuity*, 196–202 and 350–351.

20. Gearhart, "Planck, the Quantum, and the Historians," especially pp. 191–192.

21. Max Planck, *A Survey of Physical Theory*, translated by R. Jones and D. H. Williams (New York: Dover Publications, 1960), 109.

Chapter 11

1. Thomas S. Kuhn, *Black-Body Theory and the Quantum Discontinuity, 1894–1912* (Chicago: University of Chicago Press, 1987), 134–140.

2. Kuhn, *Black-Body Theory and the Quantum Discontinuity*, 140. Other historians of science dispute this argument. See, for instance, Clayton Gearhart, "Planck, the Quantum, and the Historians," *Physics in Perspective* 4 (2002): 170–215.

3. Albert Einstein, "Concerning an Heuristic Point of View toward the Emission and Transformation of Light," translated by A. B. Arons and A. B. Pippard, *American Journal of Physics* 33 (May 1965): 367–374. Originally published in *Annalen der Physik* 17 (1905): 132.

4. Einstein, "Concerning an Heuristic Point of View," 372.

5. Einstein, "Concerning an Heuristic Point of View," 373.

6. Einstein, "Concerning an Heuristic Point of View," 374.

7. It is interesting that Einstein cites Philipp Lenard and was originally supported by Johannes Stark (1874–1957), both physicists who became leaders of the "Deutsche Physik" movement that tried to discredit the "Jewish physics" of which Einstein's was a prime example. Later, Lenard and Stark, both Nobel Prize winners, became staunch supporters of the Nazi regime.

8. Albert Einstein, "Planck's Theory of Radiation and the Theory of Specific Heat," in *The Collected Papers of Albert Einstein*, vol. 2: *The Swiss Years*, translated by Anna Beck (Princeton, NJ: Princeton University Press, 1989), 214. Originally published in *Annalen der Physik* 22 (1907): 180–190.

Chapter 12

1. Albert Einstein, "On the Quantum Theory of Radiation," in *The Old Quantum Theory*, translated by D. ter Haar (Oxford: Pergamon Press, 1967), 167–183. Originally published in *Physikaliche Zeitschrift* 18, no. 121 (1917).

2. These and other quotes from Einstein's 1917 paper are from D. ter Haar's translation.

3. Daniel Kleppner identifies Einstein's quantum postulate as residing in the concept of a stationary state. See "Rereading Einstein on Radiation," *Physics Today* 58, no. 2 (2005): 30.

4. See the Wikipedia article "Spontaneous emission" in which the claim is made that "Spontaneous emission cannot be explained by classical electromagnetic theory and is fundamentally a quantum process."

The Big Ideas

1. Recall that an adiabatic invariant is only constant during an isentropic process, that is, during a quasistatic process with no friction,

internal dissipation, or heat flow. See section 2.8 for an introduction to adiabatic invariance.

2. Thomas Kuhn, *Black-Body Theory and the Quantum Discontinuity, 1894–1912* (Chicago: University of Chicago Press, 1987), 92.

3. Planck wrote the textbook: Max Planck, *Treatise on Thermodynamics*, 3rd English edition, translated by Alexander Oog (New York: Dover Publications, 1969).

4. Kuhn, *Black-Body Theory and the Quantum Discontinuity*, 47–48.

Appendix A

1. Stefan, Wien. Ber. 79. p. 391, 1879

2. The original title is "Ableitung des Stefan'schen Gesetzes, betreffend die Abhängigkeit der Wärmestrahlung von der Temperatur aus der electromagnetischen Lichttheorie; von Ludwig Boltzmann in Graz," *Annalen der Physik und Chemie* 258, no. 6 (1884): 291–294.

3. Maxwell A. Treatise on Electricity and Magnetism. Oxford, Clarendon Press vol. II Artikel 792. p. 391, 1873.

4. Krönig, Grundzüge der Theorie der Gase. Berlin bei A. W. Hain. Pogg. Ann. 99. p. 315, 1856.

5. Boltzmann. Wied. Ann. 22. p. 38, 1884.

6. $\psi(t)$ represents the quantity designated there as $4\phi(t)/cJ$, cf. p. 35.

7. Kirchhoff, Pogg. Ann. 109. p. 275. 1860. Sitzungsber. d. Berl. Acad. 1861.

8. Lecher, Wien. Ber. 85. p. 441. 1882. Wied. Ann. 17. p. 477. 1882.

Appendix B

1. The original title is "Eine neue Beziehung der Strahlung schwarzer Körper zum zweiten Hauptsatz der Wärmetheorie. Von Dr. Willy Wien in Charlottenburg. (Vorgelegt von Hrn. von Helmholtz.),"

Sitzungsberichte der Königlich Preußichen Akademie der Wissenschaften zu Berlin (1893): 55–62.

2. Wied. Ann. Bd. 32 S. 31 und 291. 1884

3. Sitzungsber. d. Berl. Akad. 1888 S. 565

Appendix C

1. This argument is adapted from C. J. Adkins, *Equilibrium Thermodynamics* (New York: McGraw-Hill, 1975), 154–159.

Index